Annual Reports in

COMPUTATIONAL CHEMISTRY

VOLUME **5**

Annual Reports in

COMPUTATIONAL CHEMISTRY

VOLUME 5

Edited by

Ralph A. Wheeler
Department of Chemistry and Biochemistry,
University of Oklahoma, 620 Parrington Oval, Room 208
Norman, Oklahoma, 73019
USA

Sponsored by the Division of Computers in Chemistry
of the American Chemical Society

ELSEVIER

Amsterdam • Boston • Heidelberg • London • New York • Oxford
Paris • San Diego • San Francisco • Singapore • Sydney • Tokyo

Elsevier
Radarweg 29, PO Box 211, 1000 AE Amsterdam, The Netherlands
Linacre House, Jordan Hill, Oxford OX2 8DP,UK
32, Jamestown Road, London NW1 7BY,UK
525 B Street, Suite 1900, San Diego, CA 92101-4495,USA
30 Corporate Drive, Suite 400, Burlington, MA 01803, USA

First edition 2009

Library of Congress Cataloging-in-Publication Data
A catalogue record for this book is available from the Library of Congress

British Library Cataloguing in Publication Data
A catalog record for this book is available from the British Library

ISBN: 978-0-444-53359-3
ISSN: 1574-1400

For information on all Elsevier publications
visit our website at elsevierdirect.com

Printed and bound in USA

09 10 11 12 13   10 9 8 7 6 5 4 3 2 1

Working together to grow
libraries in developing countries

www.elsevier.com | www.bookaid.org | www.sabre.org

ELSEVIER   BOOK AID
International   Sabre Foundation

CONTENTS

Jacques G. Amar
Department of Physics and Astronomy, University of Toledo, Toledo, OH, USA

W.F. Drew Bennett
Department of Biological Sciences, University of Calgary, Calgary, AB, Canada

Garnet Kin-Lic Chan
Department of Chemistry and Chemical Biology, Cornell University, Ithaca, NY, USA

Alan Grossfield
Department of Biochemistry and Biophysics, University of Rochester Medical Center, Rochester, NY, USA

So Hirata
Quantum Theory Project and The Center for Macromolecular Science and Engineering, Department of Chemistry and Department of Physics, University of Florida, Gainesville, FL, USA

Tingjun Hou
Functional Nano & Soft Materials Laboratory (FUNSOM), Soochow University, Suzhou, PR China

Rohit V. Pappu
Department of Biomedical Engineering, Molecular Biophysics Program, Center for Computational Biology, Washington University in St. Louis, St. Louis, MO, USA

Danny Perez
Theoretical Division, Los Alamos National Laboratory, Los Alamos, NM, USA

Yunsic Shim
Department of Physics and Astronomy, University of Toledo, Toledo, OH, USA

Toru Shiozaki
Quantum Theory Project, Department of Chemistry, University of Florida, Gainesville, FL, USA; Department of Applied Chemistry, Graduate School of Engineering, The University of Tokyo, Tokyo, Japan

Jack Simons
Chemistry Department and Henry Eyring Center for Theoretical Chemistry, University of Utah, Salt Lake City, UT, USA

D. Peter Tieleman
Department of Biological Sciences, University of Calgary, Calgary, AB, Canada

Blas P. Uberuaga
Materials Science and Technology Division, Los Alamos National Laboratory, Los Alamos, NM, USA

Edward F. Valeev
Department of Chemistry, Virginia Institute of Technology, Blacksburg, VA, USA

Andreas Vitalis
Department of Biomedical Engineering, Molecular Biophysics Program, Center for Computational Biology, Washington University in St. Louis, St. Louis, MO, USA

Arthur F. Voter
Theoretical Division, Los Alamos National Laboratory, Los Alamos, NM, USA

Junmei Wang
Department of Pharmacology, The University of Texas Southwestern Medical Center, Dallas, TX, USA

Dominika Zgid
Department of Chemistry and Chemical Biology, Cornell University, Ithaca, NY, USA

Daniel M. Zuckerman
Department of Computational Biology, University of Pittsburgh School of Medicine, Pittsburgh, PA, USA

Annual Reports in Computational Chemistry (ARCC) focuses on providing timely reviews of topics important to researchers in the field of computational chemistry. *ARCC* is published and distributed by Elsevier and is sponsored by the Division of Computers in Chemistry (COMP) of the American Chemical Society. Members in good standing of the COMP Division receive a copy of the *ARCC* as part of their member benefits. We are very pleased that previous volumes have received an enthusiastic response from our readers. The COMP Executive Committee expects to deliver future volumes of *ARCC* that build on the solid contributions in our first four volumes. To ensure that you receive future installments of this series, please join the Division as described on the COMP website at http://www.acscomp.org.

In Volume 5, our Section Editors have assembled eight contributions in four sections. Topics covered include Biological Modeling (Nathan Baker), Simulation Methodologies (Carlos Simmerling), Bioinformatics (Wei Wang), and Quantum Chemistry (Gregory Tschumper). No special emphasis was planned, but Volume 5 leans heavily toward new simulation methodologies. Although individual articles in ARCC are now indexed by the major abstracting services, we plan to continue the practice of cumulative indexing of both the current and past editions to provide easy identification of past reports.

As was the case with our previous volumes, the current volume of *Annual Reports in Computational Chemistry* has been assembled entirely by volunteers to produce a high-quality scientific publication at the lowest cost possible. The Editor and the COMP Executive Committee extend our gratitude to the many people who have given their time to make this edition of *Annual Reports in Computational Chemistry* possible. The authors of each of this year's contributions and the Section Editors have graciously dedicated significant amounts of their time to make this Volume successful. This year's edition could not have been assembled without the help of Adrian Shell of Elsevier. Thank you one and all for your hard work, your time, and your contributions.

We hope that you will find this edition to be interesting and valuable. We are actively planning the sixth volume and anticipate that it will feature a return to more applications and restore one or more previously popular sections, including Chemical Education, Materials, and/or Emerging Technologies. In addition, we are actively soliciting input from our readers about future topics, so please contact the editor to make suggestions and/or to volunteer as a contributor.

Ralph A. Wheeler
Editor

Section 1
Biological Modeling

Section Editor: Nathan Baker

Washington University School of Medicine, Department of Biochemistry and Molecular Biophysics, Center for Computational Biology, Computational and Molecular Biophysics Graduate Program, St. Louis, MO 63110, USA

CHAPTER **1**

Free Energies of Lipid–Lipid Interactions in Membranes

W.F. Drew Bennett and **D. Peter Tieleman**

Abstract   Biological membranes are lipid bilayers that function to compartmentalize cells and organelles, by forming a selectively permeable barrier. Bilayers contain a high degree of compositional and structural heterogeneity. The structure, dynamics, and phase behavior of bilayers are collective properties, which are the result of specific lipid–lipid interactions. This chapter presents an introduction to recent molecular dynamics computer simulations focused on lipid–lipid interactions, with an emphasis on free energy calculations. We discuss molecular dynamics studies focused on lipid phase behavior and mixing. We also review investigations on the free energy of pore formation in bilayers.

Keywords: phospholipid; membrane; free energy; cholesterol; pores

Department of Biological Sciences, University of Calgary, Calgary, AB, Canada

Annual Reports in Computational Chemistry, Volume 5
ISSN: 1574-1400, DOI 10.1016/S1574-1400(09)00501-5

1. INTRODUCTION

1.1 Biological and model membranes

Cellular membranes function as selective barriers and integral membrane protein scaffolds. Membranes allow the compartmentalization of cells, and individual organelles within cells, and are critical in energy transduction and cell signaling. In vivo membranes contain hundreds to thousands of lipid types, making characterization of particular lipid–lipid interactions challenging.

A remaining problem in membrane biophysics is to understand the thermodynamics of lipid mixing and, in general, membrane phase behavior and domain formation. Large-scale membrane rearrangements are critical for many biological processes, such as cell division, vesicle formation, and domain formation. There are many examples of proteins mediating membrane behavior, and in general, lipid trafficking and synthesis. More recently, examples of lipids effecting protein localization and function have been observed [1]. Since the fluid mosaic model Singer and Nicolson [2], increased structural and functional importance has been attributed to cellular lipid bilayers, with implications on the fundamental processes of life.

Biological membranes are self-aggregating lipid bilayers. They have a soft and fluid structure, as individual lipids are not covalently attached to one another, meaning intermolecular interactions and many-body effects are critical. Lipid bilayers are thin, yet they display great structural heterogeneity. Figure 1B shows a density profile of a pure DOPC bilayer. The PC head groups are zwitterionic, and the head group region has a higher density than water. Only nanometers away, the interior of the bilayer has a lower density than water, and is hydrophobic in nature. The heterogeneity of lipid membranes' structural properties is complicated by the strong dependence of the bilayer's structure on its composition and the large number of lipid types in vivo. Therefore, model systems containing only a few lipid components and no proteins must be used to simplify the problem.

The vast majority of biological membranes are in the liquid-crystalline phase. There are many experimental studies on model bilayer phase behavior [3]. Briefly, at low temperatures lipid bilayers form a gel phase, characterized by high order and rigidity and slow lateral diffusion. There is a main phase transition, as the temperature is increased, to the liquid-crystalline phase. The liquid-crystalline phase has more fluidity and fast lateral diffusion.

Phase behavior of lipid mixtures is a much more difficult problem, due to nonideal mixing of lipid components. Ideal mixing implies like and unlike lipids have the same intermolecular interactions, while nonideal mixing results from differential interactions between lipid types. If the difference is too great, the two components will phase separate. While phase separation and lateral domain formation have been observed in many experiments, we lack a molecular-level physical description of the interactions between specific lipids that cause the macroscopic behavior. The chemical potential of a lipid determines phase separation, as phase coexistence implies the chemical potential of each type of lipid is equal in all phases of the system [3,4].

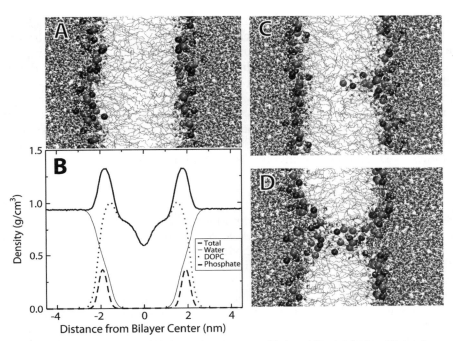

Figure 1 A: Snapshot of a DOPC bilayer during an equilibrium MD simulation. Water is denoted by small spheres, DOPC phosphorus and nitrogen are large spheres, and DOPC tails are gray lines. B: Partial density profile for a DOPC bilayer. C: A water defect present in a DOPC bilayer. D: A water pore across a DMPC bilayer. Both (C) and (D) have the same representations as (A).

Cholesterol–phospholipid interactions are of particular interest, due to the composition of eukaryotic membranes and, in particular, lipid rafts. The lipid raft hypothesis speculates cellular membranes contain domains enriched in cholesterol and sphingomyelin (SM), which are crucial for cellular signaling and lipid trafficking [5–7]. According to this hypothesis, rafts would be in the liquid-ordered phase, while the rest of the membrane would be in the liquid-disordered phase. Most evidence suggests rafts are small (10–200 nm diameter), highly dynamic, and heterogeneous [8]. Keller and coworkers have shown both giant plasma vesicles and model vesicles near their respective critical points have critical fluctuations with correlation lengths similar to the speculated diameter of lipid rafts [9,10]. A recent study employing the new spectroscopic technique of stimulated emission depletion (STED) far-field fluorescence nanoscopy provided a direct observation of cholesterol complexes in living cells [11].

There is a large cholesterol concentration gradient in cells from 0–5 mol% in the ER membrane to 25–40 mol% in the plasma membrane [12]. Cholesterol has a condensing effect on liquid-crystalline bilayers, causing increased rigidity and thickness [13]. At high concentrations, cholesterol induces an intermediate liquid-ordered phase between the gel and liquid-disordered phases [13]. A number of experimental techniques have studied the phase diagram for DPPC–cholesterol mixtures, but there is still debate about whether it contains a liquid-ordered–liquid-disordered phase coexistence region [3]. More recently, three-component phase diagrams of cholesterol–a high melting temperature phospholipid–a low melting temperature phospholipid (e.g., cholesterol–DOPC–DPPC) have been determined [3]. These diagrams do display regions of liquid-ordered–liquid-disordered coexistence, and domains can be visualized in vitro. The microscopic interactions responsible for the anomalous behavior of cholesterol–phospholipid mixtures remain unresolved. Section 2 of this chapter looks at molecular dynamics (MD) studies that involve lipid mixing and phase behavior.

1.2 Pores and defects

Pore formation in membranes is an important process in biology and biotechnology. Many antimicrobial peptides induce pore formation in cells, which causes dissipation of the cell electrochemical gradient [14]. Understanding the mechanism and energetics of antimicrobial peptides is necessary for effective clinical use. Cationic penetrating peptides (CPPs) have been used to translocate biological macromolecules across cellular membranes, although the detailed mechanism remains controversial [15]. Experimental evidence supports both an energy-independent, nonendocytotic mechanism and an endocytotic mechanism [16]. Electric fields have been used extensively to induce pore formation in membranes, such as to introduce DNA or other foreign molecules into cells. Recently, high-energy ultra-short laser pulses have been used to induce apoptosis by permeabilizing membranes, with potential therapeutic applications [17]. Thermal fluctuations can cause spontaneous pore formation, important for passive lipid flip-flop and ion permeation, but are rare due to their high free-energy barriers [18]. Lipid flip-flop is an important process in general lipid synthesis and trafficking. As most lipids are synthesized on the cytoplasmic side of biogenic membranes, such as the endoplasmic reticulum, flip-flop is required for even growth of the membrane, and for lipids destined for the extracellular leaflet of the plasma membrane [19]. The endoplasmic reticulum displays a symmetric distribution of lipids, while the plasma membrane maintains an asymmetric distribution [19]. Disruption of the asymmetry can lead to apoptosis [20]. While proteins mediate the process of cellular lipid trafficking, passive flip-flop is still likely physiologically important and not well understood. Finally, pore formation is likely an intermediate in large-scale rearrangements of membranes, such as during vesicle fusion.

Lipid membranes are quite deformable, allowing water and head groups into their interiors when perturbed. A "water defect" is shown in Figure 1C, where water and lipid head groups enter the hydrophobic interior of only one of the bilayer leaflets. Figure 1D shows a "water pore," where both leaflets are perturbed. At the molecular level, pore and defect formation are directly related to specific lipid–lipid interactions. It is important to understand the free energy required for pore formation in membranes and the effect of lipid composition on the process. In Section 3 of this chapter, we review recent MD studies of the thermodynamics of pore formation.

1.3 MD simulations of membranes

MD simulations of model membrane systems have provided a unique view of lipid interactions at a molecular level of resolution [21]. Due to the inherent fluidity and heterogeneity in lipid membranes, computer simulation is an attractive tool. MD simulations allow us to obtain structural, dynamic, and energetic information about model lipid membranes. Comparing calculated structural properties from our simulations to experimental values, such as areas and volumes per lipid, and electron density profiles, allows validation of our models. With molecular resolution, we are able to probe lipid–lipid interactions at a level difficult to achieve experimentally.

The parameterization of lipid force fields is challenging. There is a lack of experimental data to compare to when parameterizing or validating a force field. Current force fields are judged by structural data such as density profiles and the area per lipid. It has been demonstrated that many measurable properties, such as the area per lipid, are not sensitive enough to be reliable validators or targets in force field development [22]. One interesting point is that a multitude of experimental data exists of lipid bilayer phase behavior [3], although these data have not been exploited for force field development yet.

The other major limitation of membrane simulations is the time and length scale we are able to simulate. We are currently able to reach a microsecond, but tens to hundreds of nanosecond simulations are more common, especially in free energy calculations. The slow diffusion of lipids means we are not able to observe many biologically interesting phenomena using equilibrium simulations. For example, we would not observe pore formation in an unperturbed bilayer system during an equilibrium simulation, and even pore dissipation is at the limits of current computational accessibility.

There has been considerable effort in developing coarse-grained models for lipids to overcome the time- and length-scale limitations. By reducing the number of degrees of freedom and smoothing the energy landscape, larger systems can be simulated for much longer times. In this chapter, we focus on atomistic simulations, but mention a few coarse-grained simulations. One example of a coarse-grained model that is particularly accurate for lipids is the MARTINI model [23]. The MARTINI model effectively replaces three to four heavy atoms with a bead, parameterized to reproduce condensed-phase thermodynamic data of small molecules [23]. The MARTINI model has been used to investigate many biological processes, such as lung surfactant collapse [24], nanoparticle permeation in bilayers [25], large domain motion of integral membrane proteins [26], vesicle fusion [27,28], and lateral domain formation in membranes [29].

2. LIPID MIXING AND PHASE BEHAVIOR

Lipid phase transitions are often slow on the timescale accessible to simulations, such as the gel to liquid-crystalline phase transition. Using the MARTINI model, liquid-crystalline to gel phase transitions and domain formation have been observed in equilibrium simulations [29,30]. For mixtures, due to the slow diffusion of lipids, lateral domain formation has not been observed in atomistic simulations. Obtaining properly mixed model bilayers using equilibrium simulations is difficult, and the results are hard to evaluate.

We must turn to alternative methods to investigate lipid mixing and phase behavior in MD simulations. Free energies directly determine lipid mixing and phase behavior, and in general, bilayer structure. Determining free energies from MD simulations is not straightforward, as it usually involves defining a particular reaction coordinate and sampling all relevant degrees of freedom orthogonal to the reaction coordinate. The slow diffusion of lipids makes adequate sampling computationally expensive.

2.1 Cholesterol–phospholipid interactions

MD simulations have provided a unique molecular description of cholesterol–phospholipid interactions [31]. Atomistic simulations have succeeded in reproducing the condensing effect of cholesterol on phospholipid bilayers [32–34]. With atomistic detail, many properties can be determined, such as the effect of cholesterol on lipid chain ordering or on hydrogen bond formation. Other simulations have focused on the interaction of cholesterol and SM [35–37]. Aittoniemi et al. [38] showed that hydrogen bonding alone cannot explain the preferential interaction between cholesterol and SM compared to cholesterol and POPC.

The interaction of cholesterol with polyunsaturated lipid tails has also been investigated using MD. Pitman et al. [39] showed cholesterol interacts preferentially with the saturated tail of SDPC (18:0–22:6-PC) compared to the unsaturated tail. Using the MARTINI model, Marrink and coworkers [40] showed cholesterol has a preference for the interior of a DAPC bilayer (di20:4-PC), which agreed well with neutron-scattering data [41]. In equilibrium MARTINI simulations, Risselada and Marrink [29] observed phase separation of a DLPC (di18:2-PC), DPPC (di16:0-PC), and cholesterol mixture. The molecular resolution allowed detailed analysis of the liquid-ordered and -disordered domains. The preference of cholesterol for saturated acyl tails compared to polyunsaturated tails has been demonstrated experimentally, using cyclodextrin cholesterol extraction [42] and fluorescence microscopy [43].

Ternary mixtures involving cholesterol have been used to simulate putative lipid rafts. Niemela and coworkers [44] provided an in-depth structural and dynamic comparison of "raft" and "non-raft" bilayers. By determining the elastic properties and lateral pressure profiles of the bilayers, they provided a molecular-level picture of lipid rafts. Using the pressure profile, they estimated the free energy associated with the bilayer volume changes caused by integral membrane protein conformational changes [44]. This provided a possible mechanism for lipid raft effect on integral membrane protein function, with implications on cellular signaling. Pandit et al. [45] investigated a putative liquid-ordered domain embedded in a DOPC bilayer and found that the liquid-ordered patch was thicker.

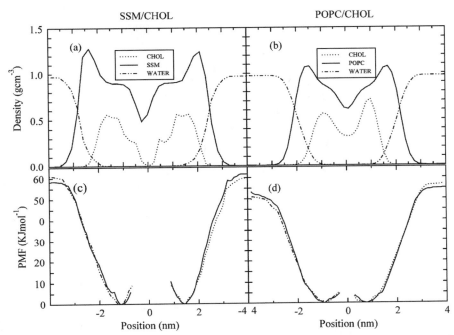

Figure 2 PMFs and density profiles for cholesterol in a SSM and POPC bilayer. a, b: Partial density profiles for the two bilayer systems. The cholesterol density was multiplied by a factor of 20 for visualization. c, d: PMFs for cholesterol transfer from equilibrium of the respective bilayer to bulk water. The center of mass of the cholesterol molecule was restrained with respect to the center of the bilayer. Reprinted with permission from ref. 46. Copyright 2009 American Chemical Society.

Only recently have free energies of cholesterol–lipid interactions been determined using atomistic MD simulations. Zhang and coworkers [46] determined a potential of mean force (PMF) using umbrella sampling for transferring cholesterol from a POPC bilayer to water and also from a SM bilayer to water. Figure 2 shows the PMFs, which have steep slopes as the cholesterol is transferred from its equilibrium position into bulk water. They showed that the transfer from POPC to SM was favorable by 2–3 kT. The favorable free energy of transfer was decomposed by determining the PMF at 319 and 329 K, and was shown to be entropically unfavorable, and therefore enthalpically driven. Their results qualitatively agreed with calorimetric data, which showed the process was exothermic [47]. In a previous study, the potential energies and nearest-neighbor interactions between cholesterol–POPC and cholesterol–SSM mixtures were similar, suggesting the transfer would be entropically driven [36]. This illustrates the importance of free energies as opposed to other properties that lack an entropic component.

Using umbrella sampling with both atomistic and MARTINI simulations, we calculated PMFs for transferring cholesterol from water to the center of a series of lipid bilayers [48]. Figure 3A shows cholesterol PMFs in a DAPC (di20:4-PC)

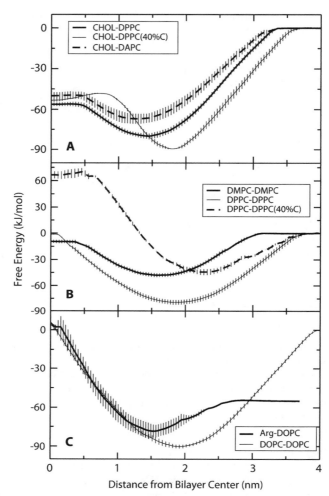

Figure 3 PMFs for lipids partitioning into different bilayers. A: Cholesterol partitioning from water to the center of a pure DPPC bilayer, a DPPC bilayer with 40 mol% cholesterol, and a DAPC bilayer. The hydroxyl group of the cholesterol was restrained during the umbrella sampling. B: PMFs for DMPC in a DMPC bilayer, DPPC in a pure DPPC bilayer, and DPPC in a DPPC bilayer with 40 mol% cholesterol. The phosphate of the PC lipid was restrained for the umbrella sampling. C: PMFs for an analog of arginine in a DOPC bilayer, and a DOPC lipid in a DOPC bilayer. The arginine PMF has been shifted to illustrate the similarity in the slope of the PMFs.

bilayer, a DPPC bilayer, and a DPPC bilayer with 40 mol% cholesterol. The free energy difference between cholesterol's equilibrium position in the bilayer and bulk water (ΔG_{desorb}) is the excess chemical potential of cholesterol in the bilayer compared to water. Cholesterol had the highest affinity for rigid, ordered saturated DPPC bilayers with 40 mol% cholesterol ($\Delta G_{desorb} = 89 \, \text{kJ/mol}$). At infinite dilution, ΔG_{desorb} was 80 kJ/mol in a DPPC bilayer. The lowest affinity of cholesterol was for a polyunsaturated DAPC (di20:4-PC) bilayer ($\Delta G_{desorb} = 68$ kJ/mol). These results support the above-mentioned preference of cholesterol for saturated lipid tails. The gradient in ΔG_{desorb} in a DPPC bilayer, as the cholesterol content is increased, suggests nonideal mixing of the cholesterol and DPPC. We determined PMFs at 323, 333, and 343 K to decompose the free energy at 333 K, and showed that desorption was entropically favorable.

2.2 Phospholipid–phospholipid interactions

Using umbrella sampling, Tieleman and Marrink [18] determined a PMF for transferring a DPPC lipid from water to the center of a DPPC bilayer (Figure 3B). The DPPC PMF has a deep minimum at its equilibrium position and a steep slope in free energy as it moved into bulk water. The free energy of desorption (ΔG_{desorb}) was 80 kJ/mol, and is directly related to its excess chemical potential in the bilayer compared to water.

Recently, Kindt and coworkers have developed a mixed MC/MD method, which allows sampling of the semigrand-canonical ensemble [49]. In this scheme, lipid species are exchanged with one another to ensure the activity ratio of the two species is maintained throughout the simulation. After every time step a single lipid attempts to exchange to the other type, with a MC-based criterion. Various systems were simulated at different constant activity ratios and the effect on the bilayer composition was monitored, as well as the structural properties of the mixed bilayer. By allowing lipids to exchange, efficient mixing is obtained on a nanosecond timescale. For example, the mixing of DPPC and DLPC (di12:0-PC) was investigated using this method [49]. From the lateral distribution of the lipids, near-ideal mixing was observed, although a small unfavorable free energy of mixing was inferred from the effect of the activity ratio on the composition. This method has been used to study the effect of bilayer curvature on the mixing of DMPC (di14:0-PC) and the short-chain lipids DDPC (di10:0-PC) and HMPC (14:0-6:0-PC) [50]. It was shown that the shorter lipids were enriched on the edge of a bilayer ribbon in agreement with previous CG simulations, experiments, and theoretical work [50]. DSPC (di18:0-PC) and DMPC (di14:0-PC) mixing in both the liquid-crystalline and gel phases was investigated by Coppock and Kindt [51]. In the liquid-crystalline phase, only a slight deviation from ideal mixing was observed, in contrast to the gel phase, which displayed largely nonideal mixing. A limitation of this model is that only very similar lipids can be investigated; for example, DPPC to cholesterol exchanges would not be possible. The authors speculate that differences of more than eight carbons per lipid would be problematic [49].

Rodriguez et al. [52] determined the association free energy of DPPS in a DPPC bilayer, both in the presence and in the absence of calcium ions. They used a dual topology technique with a PC-PS hybrid, and used both thermodynamic integration and free energy perturbation to determine the free energy of DPPS association in a DPPC bilayer. They found that PS–PS association is only favorable in the presence of calcium, which has direct implications for blood coagulation.

Using a similar approach as in ref. [18], Sapay et al. calculated ΔG_{desorb} for a systematic series of lipids. We found ΔG_{desorb} for DLPC, DMPC, DPPC, POPC, and DOPC are equal to 39, 48, 80, 85, and 90 kJ/mol [53]. This qualitatively agrees with the critical micelle concentrations for DLPC, DMPC, and DPPC. This trend makes sense intuitively: as the number of carbons in the acyl tail increases, the exposure to water becomes more unfavorable.

We have also undertaken MD simulations to examine the effect of cholesterol content on the thermodynamics of DPPC desorption [54]. We found that DPPC had a lower affinity for bilayers with high cholesterol content (Figure 3B). This suggests that while cholesterol prefers to interact with saturated lipid tails, the saturated tails might not prefer to interact with cholesterol. It would be interesting to repeat this study on unsaturated lipid tails.

3. PORES, DEFECTS, AND FLIP-FLOP

MD simulations have investigated pore and defect formation in membranes, both directly and indirectly, using a variety of techniques. A common structure of membrane pores has emerged from these studies: pores have a toroidal shape, with water and lipid head groups penetrating into the hydrophobic interior of the bilayer (Figure 1). These pores have been characterized as disordered toroidal pores due to their structural heterogeneity and fluidity [55]. Only recently have free energy calculations involving pores been undertaken for atomistic simulations, due to the slow diffusion of lipids and the large structural rearrangements involved in pore formation.

3.1 Non-free energy pore simulations

Antimicrobial peptides and CPPs have been shown to induce pore formation in bilayers during equilibrium atomistic MD simulations. Herce and Garcia [56] showed that the HIV-1 Tat peptide translocated across a bilayer through a large pore. Marrink and coworkers have observed pore formation caused by both magainin [57] and melittin [55]. Charged arginines on the voltage-sensing S1–S4 domains have also been shown to cause water defects in membranes [58–60].

MD simulations have been used to investigate electroporation of lipid bilayers by directly applying an electric field to a bilayer [61–64]. Alternatively, an ion imbalance across a bilayer can be used to create an electrochemical gradient, also leading to pore formation [65–68]. The presence of dimethylsulfoxide (DMSO) in high concentrations has been shown to increase the permeability of bilayers and induce pore formation [69–71]. Mechanical stress has been used to induce pore formation in MD simulations [72,73]. By applying a surface tension to DPPC bilayers, it was shown that pores with a critical radius of 0.7 nm could be stabilized, while larger pores caused membrane rupture [73]. From the critical threshold tension, the line tension of the bilayer was found to be in good agreement with experimental values.

3.2 Small molecule partitioning

We mention a few studies on polar and charged molecule–lipid interactions, as their permeation involves defect and pore formation, which involves lipid–lipid interactions. A detailed examination of small molecule–lipid interactions is beyond the scope of this chapter [74]. We have shown that a large component of the free energy of small polar or charged molecule partitioning into lipid bilayers is due to the cost of forming a defect [75].

MacCallum et al. [75] determined PMFs for 17 of the 20 amino acid side-chain analogs in a DOPC bilayer. The free energies of transfer compared well to various experimentally determined hydrophobicity scales. Of interest to our topic, the polar and charged amino acids caused significant water defects when placed in the hydrophobic interior of the bilayer. Water and DOPC head groups moved into the interior to prevent the desolvation of the side chain. There was an energetic balance between desolvation and water defect formation. Moving the side chain further into the bilayer caused the water defect to get larger, and translated into a steep slope in the PMF. At a critical distance from the bilayer center, the water defect dissipated and the PMFs flattened. The steep slopes of the PMFs were similar for all the polar side chains and charged side chains, which suggests that the cost of water defect formation dominates the PMF. We note that the slope for transferring a DOPC lipid to the bilayer center was also similar to the charged side chains (Figure 3C) [53]. The distance from the bilayer center at which the defect broke — the distance when desolvation was cheaper than defect formation — depended on the side chain. For the charged side chains, and the zwitterionic DOPC, a water defect was present, even at the center of the membrane. Similar studies on arginine partitioning, using different methods and force fields, have yielded similar results [76,77]. We have shown that transferring an arginine into a bilayer that already contains a water defect is energetically inexpensive (12 kJ/mol) [MacCallum et al., in preparation]. This supports our assertion that defect formation dominates the free energy cost for charged and polar molecules partitioning in lipid bilayers.

3.3 Free energies of pore formation

Using a simplified coarse-grained model, Tolpekina et al. [78] have investigated pore formation by calculating a PMF as a function of pore radius. They found that pore formation in their coarse-grained model cost 15–20 kT. The PMF did not show a barrier to pore closure, which had been suggested from experimental evidence, although their results did not rule it out. In another study, the same group determined a pore phase diagram by stretching a coarse-grained bilayer and determining regions of stability, metastability, and instability [79]. The line tension coefficient along the pore edge was determined using three different methods, all in good agreement with the experimental value.

Wohlert et al. [80] determined the free energy of pore formation in an atomistic DPPC bilayer using the pore radius as the reaction coordinate. From this reaction coordinate, they were able to derive the free energy of pore

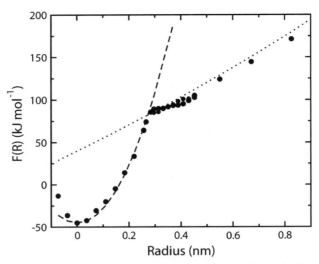

Figure 4 The free energy of pore formation in a DPPC bilayer. The dashed line is a quadratic function, while the dotted line is a fit to a model of pore expansion with a line tension of 40 pN, and is close to linear (Adapted from ref. 78 courtesy of O. Edholm).

formation and pore expansion (Figure 4). For pore radii of less than ~ 0.3 nm, the PMF had a quadratic shape. At larger pore radii, or pore expansion, the PMF was linear. They found the free energy for pore formation was 75–100 kJ/mol.

Using a similar approach, Notman et al. [81], determined the free energy for pore formation in bilayers composed of ceramide, as a model for the stratum corneum of the skin, both in the presence and in the absence of DMSO. Without DMSO, the bilayer was in the gel phase, and interestingly, a hydrophobic pore was observed with a high free-energy barrier (~ 60 kJ/mol). In the presence of DMSO, the bilayer was more fluid, and the more typical hydrophilic pore was observed, with a much smaller activation energy of 20 kJ/mol. This work provided a thermodynamic and structural explanation for the enhanced permeability of skin by DMSO.

3.4 Lipid flip-flop

The free energy for pore formation was obtained from the DPPC PMF calculated by Tieleman and Marrink [18]. A pore was observed when the phosphate of a DPPC molecule was restrained at the center of a DPPC bilayer (Figure 3B). The free energy for flip-flop and pore formation was 80 kJ/mol. This is in good agreement with the free energy cost of pore formation in a DPPC bilayer determined by Wohlert et al. [80]. It appears that the cost for an initial pore is the cost of placing a single lipid head group in the center of the bilayer, at least for DPPC. With the assumption that this state is the transition state for lipid flip-flop combined with calculated diffusion rates of lipids across the bilayer using preformed pores, flip-flop rates can be calculated. The timescale of DPPC flip-flop was estimated to be 1–30 h, well within the range of experimental measurements [18].

Using preformed pores in a DMPC bilayer, Gurtovenko and Vattulainen [82] investigated the translocation of DMPC across a pore. It was shown that multiple lipids diffused across the pore before it dissipated, providing support for pore-mediated flip-flop as mechanism for passive flip-flop. The timescale for pore dissipation was found to be 35–200 ns, at the limits of current computational capability for equilibrium simulations.

From the PMFs of Sapay et al., the free energy barrier for flip-flop was obtained for a series of PC lipid bilayers: ΔG_{flop} for DLPC, DMPC, DPPC, POPC, and DOPC are equal to 16, 40, 80, 89, and 94 kJ/mol, respectively [53]. They found a large difference in the free energy barriers for flip-flop. In general, the free energy barrier for flip-flop increased as the bilayer became thicker. For DLPC and DMPC, they observed water pore formation when the pulled lipid was near the bilayer center. Once a pore formed, the PMF flattened, indicating the lipids could then freely diffuse across the pore (Figure 3B). For the thicker POPC and DOPC bilayers, pore formation was not observed, and only a water defect was present at the bilayer center.

The effect of cholesterol on DPPC partitioning was also investigated using MD simulations and umbrella sampling [54]. PMFs for transferring DPPC from water to the center of DPPC bilayers containing 20 and 40 mol% cholesterol were determined by umbrella sampling (Figure 3B). We found that cholesterol had a concentration-dependent effect on the free energy required for DPPC flip-flop. For the 20 mol% cholesterol bilayer, DPPC formed a water defect, while no defect was present for the 40 mol% bilayer. Instead, a small number of water molecules were pulled into the 40 mol% bilayer to solvate the DPPC head group. Cholesterol was shown to impede the formation of water pores and defects, thereby reducing the rate of DPPC flip-flop by orders of magnitude.

Cholesterol flip-flop is important for general cholesterol trafficking as well as for lateral domain formation. Using methyl-β-cyclodextrin as a cholesterol acceptor, Steck et al. [83] estimated the halftime for cholesterol flip-flop in a human red cell bilayer to be <1 s.

By replacing the hydroxyl head group of cholesterol with a ketone, Rog et al. [84] observed flip-flop in atomistic equilibrium simulations on a nanosecond timescale. This suggests the barrier for ketosterone flip-flop is much reduced compared to cholesterol. Using combined atomistic simulations and X-ray scattering, Kucerka et al. [85] investigated cholesterol distribution in short-chain (di14:1-PC) and long-chain (di22:1-PC) lipid bilayers. In the short-chain bilayer, cholesterol was found to occasionally reside in the hydrophobic interior, stabilized by a hydrogen-bonded network of water. Using the MARTINI model, cholesterol flip-flop has been observed in equilibrium simulations [29,40]. Fast cholesterol flip-flop and appreciable residence times in the hydrophobic interior of polyunsaturated bilayers were observed [40], in agreement with neutron-scattering data [41]. Risselada and Marrink [29] showed cholesterol has a high rate of flip-flop in liquid-disordered domain, and low rate of flip-flop in the liquid-ordered domain.

The free energy for cholesterol flip-flop in a series of lipid membranes was obtained from umbrella sampling using atomistic and MARTINI MD simulations (Figure 3A) [48]. The rate of flip-flop was estimated using the free energy at the bilayer center compared to equilibrium and the time required for cholesterol to move from the bilayer center back to equilibrium. Using the MARTINI model, cholesterol flip-flop was directly observed during equilibrium simulations, and compared to the rate estimated from umbrella sampling. During cholesterol flip-flop, water pore formation was not observed. Similar to the polar side-chain PMFs discussed above, a small water defect formed as cholesterol was moved into the bilayer interior, but at a critical distance from the bilayer center, the defect dissipated and the PMF flattened (Figure 3A). We found that the rate of cholesterol flip-flop is fast, but dependent on the structure and composition of the bilayer. For example, in a polyunsaturated DAPC (di20:4-PC) bilayer, flip-flop would occur on the submicrosecond timescale, and on the second range in a saturated DPPC bilayer with a high concentration of cholesterol.

4. CONCLUSIONS

MD simulations have provided a unique molecular-level view of lipid–lipid interactions, as well as insight into the thermodynamic driving forces of lipid mixing and pore formation. Bilayer phase behavior and mixing are directly related to specific lipid–lipid interactions. From simulations, valuable insight into the physicochemical nature of cholesterol–phospholipid interactions has been gained. Recently, a range of methods have been used to calculate free energies of cholesterol–phospholipid and phospholipid–phospholipid interactions.

Pore formation in biological membranes has important implications. Bilayers are soft and deformable, and the cost of defect and pore formation is related to specific lipid–lipid interactions. MD simulations have provided a structural picture of pores from a variety of techniques. Hydrophilic pores have a disordered toroidal shape with water and head groups penetrating into the bilayer interior. The stability of pores and the free energy of pore formation depend on the pore radius. As well, the free energy required for pore formation depends on the structure and composition of the bilayer. More ordered and rigid bilayers have a higher free energy barrier than thinner and more fluid bilayers. Lipid bilayers are able to bend but not break, which enables many interesting biological and potentially therapeutic processes.

Due to the high computational expense for calculating free energies in lipid bilayer systems, most of the studies we have reviewed are quite recent. In the future, we anticipate an increase in atomistic lipid simulations focused on free energy calculations. With advances in computer power and algorithms for MD, free energy calculations will become routine. It may become possible to investigate bilayer phase diagrams using MD simulations, and possibly parameterize new force fields based on the thermodynamics of phase behavior. Additionally, advances in experimental techniques may provide a new level of resolution, which we will be able to compare our results against, and gain insight into new directions of study.

ABBREVIATIONS

DPPC	dipalmitoylphosphatidylcholine
DPPS	dipalmitoylphosphatidylserine
POPC	1-palmitoyl-2-oleoyl-phosphatidylcholine
DOPC	dioleoylphosphatidylcholine
SM	sphingomyelin
DAPC	diarachidonylphosphatidylcholine
DLPC	dilauroylphosphatidylcholine
DMPC	dimyristoylphosphatidylcholine
DMSO	dimethylsulfoxide
CPP	cationic penetrating peptide
MD	molecular dynamics
MC	Monte Carlo
PMF	potential of mean force

ACKNOWLEDGMENTS

This work is supported by the Natural Sciences and Engineering Research Council (Canada). W.F.D. Bennett is supported by studentships from NSERC and AHFMR. D.P. Tieleman is an Alberta Heritage Foundation for Medical Research Senior Scholar and Canadian Institutes of Health Research New Investigator.

REFERENCES

1. Sprong, H., van der Sluijs, P., van Meer, G. How proteins move lipids and lipids move proteins. Nat. Rev. Mol. Cell Biol. 2001, 2, 504–13.
2. Singer, S.J., Nicolson, G.L. The fluid mosaic model of the structure of cell membranes. Science 1972, 175, 720–31.
3. Veatch, S.L., Keller, S.L. Seeing spots: complex phase behavior in simple membranes. Biochim. Biophys. Acta 2005, 1746, 172–85.
4. Lee, A.G. Lipid phase-transitions and phase-diagrams .2. Mixtures involving lipids. Biochim. Biophys. Acta 1977, 472, 285–344.
5. Brown, D.A., London, E. Structure of detergent-resistant membrane domains: does phase separation occur in biological membranes? Biochem. Biophys. Res. Commun. 1997, 240, 1–7.
6. Edidin, M. The state of lipid rafts: from model membranes to cells. Ann. Rev. Biophys. Biomol. Struct. 2003, 32, 257–83.
7. Simons, K., Ikonen, E. Functional rafts in cell membranes. Nature 1997, 387, 569–72.
8. Pike, L.J. Rafts defined: a report on the Keystone Symposium on Lipid Rafts and Cell Function. J. Lipid Res. 2006, 47, 1597–8.
9. Veatch, S.L., Cicuta, P., Sengupta, P., Honerkamp-Smith, A., Holowka, D., Baird, B. Critical fluctuations in plasma membrane vesicles. ACS Chem. Biol. 2008, 3, 287–93.
10. Veatch, S.L., Soubias, O., Keller, S.L., Gawrisch, K. Critical fluctuations in domain-forming lipid mixtures. Proc. Natl. Acad. Sci. U.S.A. 2007, 104, 17650–5.

11. Eggeling, C., Ringemann, C., Medda, R., Schwarzmann, G., Sandhoff, K., Polyakova, S., Belov, V.N. et al. Direct observation of the nanoscale dynamics of membrane lipids in a living cell. Nature 2009, 457, 1159–62.
12. Ikonen, E. Cellular cholesterol trafficking and compartmentalization. Nat. Rev. Mol. Cell. Biol. 2008, 9, 125–38.
13. McMullen, T.P.W., Lewis, R.N.A.H., McElhaney, R.N. Cholesterol-phospholipid interactions, the liquid-ordered phase and lipid rafts in model and biological membranes. Curr. Opin. Colloid Interface Sci. 2004, 8, 459–68.
14. Epand, R.M., Vogel, H.J. Diversity of antimicrobial peptides and their mechanisms of action. Biochim. Biophys. Acta Biomem. 1999, 1462, 11–28.
15. Duchardt, F., Fotin-Mleczek, M., Schwarz, H., Fischer, R., Brock, R. A comprehensive model for the cellular uptake of cationic cell-penetrating peptides. Traffic 2007, 8, 848–66.
16. Henriques, S.T., Melo, M.N., Castanho, M.A.R.B. Cell-penetrating peptides and antimicrobial peptides: how different are they? Biochem. J. 2006, 399, 1–7.
17. Vernier, P.T., Sun, Y.H., Marcu, L., Craft, C.M., Gundersen, M.A. Nanoelectropulse-induced phosphatidylserine translocation. Biophys. J. 2004, 86, 4040–48.
18. Tieleman, D.P., Marrink, S.J. Lipids out of equilibrium: energetics of desorption and pore mediated flip-flop. J. Am. Chem. Soc. 2006, 128, 12462–7.
19. van Meer, G., Voelker, D.R., Feigenson, G.W. Membrane lipids: where they are and how they behave. Nat. Rev. Mol. Cell Biol. 2008, 9, 112–24.
20. Daleke, D.L. Regulation of transbilayer plasma membrane phospholipid asymmetry. J. Lipid Res. 2003, 44, 233–42.
21. Tieleman, D.P., Marrink, S.J., Berendsen, H.J.C. A computer perspective of membranes: molecular dynamics studies of lipid bilayer systems. Biochim. Biophys. Acta Biomem. 1997, 1331, 235–70.
22. Anezo, C., de Vries, A.H., Holtje, H.D., Tieleman, D.P., Marrink, S.J. Methodological issues in lipid bilayer simulations. J. Phys. Chem. B 2003, 107, 9424–33.
23. Marrink, S.J., Risselada, H.J., Yefimov, S., Tieleman, D.P., de Vries, A.H. The MARTINI force field: coarse grained model for biomolecular simulations. J. Phys. Chem. B 2007, 111, 7812–24.
24. Baoukina, S., Monticelli, L., Risselada, H.J., Marrink, S.J., Tieleman, D.P. The molecular mechanism of lipid monolayer collapse. Proc. Natl. Acad. Sci. U.S.A. 2008, 105, 10803–08.
25. Wong-Ekkabut, J., Baoukina, S., Triampo, W., Tang, I.M., Tieleman, D.P., Monticelli, L. Computer simulation study of fullerene translocation through lipid membranes. Nat. Nanotechnol. 2008, 3, 363–8.
26. Treptow, W., Marrink, S.J., Tarek, M. Gating motions in voltage-gated potassium channels revealed by coarse-grained molecular dynamics simulations. J. Phys. Chem. B 2008, 112, 3277–82.
27. Kasson, P.M., Pande, V.S. Control of membrane fusion mechanism by lipid composition: predictions from ensemble molecular dynamics. PLoS Comput. Biol. 2007, 3, e220.
28. Marrink, S.J., Mark, A.E. The mechanism of vesicle fusion as revealed by molecular dynamics simulations. J. Am. Chem. Soc. 2003, 125, 11144–5.
29. Risselada, H.J., Marrink, S.J. The molecular face of lipid rafts in model membranes. Proc. Natl. Acad. Sci. U.S.A. 2008, 105, 17367–72.
30. Faller, R., Marrink, S.J. Simulation of domain formation in DLPC-DSPC mixed bilayers. Langmuir 2004, 20, 7686–93.
31. Berkowitz, M.L. Detailed molecular dynamics simulations of model biological membranes containing cholesterol. Biochim. Biophys. Acta 2009, 1788, 86–96.
32. Hofsass, C., Lindahl, E., Edholm, O. Molecular dynamics simulations of phospholipid bilayers with cholesterol. Biophys. J. 2003, 84, 2192–206.
33. Chiu, S.W., Jakobsson, E., Mashl, R.J., Scott, H.L. Cholesterol-induced modifications in lipid bilayers: a simulation study. Biophys. J. 2002, 83, 1842–53.
34. Falck, E., Patra, M., Karttunen, M., Hyvonen, M.T., Vattulainen, I. Lessons of slicing membranes: interplay of packing, free area, and lateral diffusion in phospholipid/cholesterol bilayers. Biophys. J. 2004, 87, 1076–91.
35. Rog, T., Pasenkiewicz-Gierula, M. Cholesterol-sphingomyelin interactions: a molecular dynamics simulation study. Biophys. J. 2006, 91, 3756–67.

36. Zhang, Z., Bhide, S.Y., Berkowitz, M.L. Molecular dynamics simulations of bilayers containing mixtures of sphingomyelin with cholesterol and phosphatidylcholine with cholesterol. J. Phys. Chem. B 2007, 111, 12888–97.
37. Khelashvili, G.A., Scott, H.L. Combined Monte Carlo and molecular dynamics simulation of hydrated 18:0 sphingomyelin-cholesterol lipid bilayers. J. Chem. Phys. 2004, 120, 9841–7.
38. Aittoniemi, J., Niemela, P.S., Hyvonen, M.T., Karttunen, M., Vattulainen, I. Insight into the putative specific interactions between cholesterol, sphingomyelin, and palmitoyl-oleoyl phosphatidylcholine. Biophys. J. 2007, 92, 1125–37.
39. Pitman, M.C., Suits, F., Mackerell, A.D. Jr. , Feller, S.E. Molecular-level organization of saturated and polyunsaturated fatty acids in a phosphatidylcholine bilayer containing cholesterol. Biochem. 2004, 43, 15318–28.
40. Marrink, S.J., de Vries, A.H., Harroun, T.A., Katsaras, J., Wassall, S.R. Cholesterol shows preference for the interior of polyunsaturated lipid membranes. J. Am. Chem. Soc. 2008, 130, 10–1.
41. Harroun, T.A., Katsaras, J., Wassall, S.R. Cholesterol hydroxyl group is found to reside in the center of a polyunsaturated lipid membrane. Biochem. 2006, 45, 1227–33.
42. Leventis, R., Silvius, J.R. Use of cyclodextrins to monitor transbilayer movement and differential lipid affinities of cholesterol. Biophys. J. 2001, 81, 2257–67.
43. Mitchell, D.C., Litman, B.J. Effect of cholesterol on molecular order and dynamics in highly polyunsaturated phospholipid bilayers. Biophys. J. 1998, 75, 896–908.
44. Niemela, P.S., Ollila, S., Hyvonen, M.T., Karttunen, M., Vattulainen, I. Assessing the nature of lipid raft membranes. PloS Comput. Biol. 2007, 3, 304–12.
45. Pandit, S.A., Vasudevan, S., Chiu, S.W., Mashl, R.J., Jakobsson, E., Scott, H.L. Sphingomyelin-cholesterol domains in phospholipid membranes: atomistic simulation. Biophys. J. 2004, 87, 1092–100.
46. Zhang, Z., Lu, L., Berkowitz, M.L. Energetics of cholesterol transfer between lipid bilayers. J. Phys. Chem. B 2008, 112, 3807–11.
47. Tsamaloukas, A., Szadkowska, H., Heerklotz, H. Thermodynamic comparison of the interactions of cholesterol with unsaturated phospholipid and sphingomyelins. Biophys. J. 2006, 90, 4479–87.
48. Bennet, W.F.D., MacCallum, J.L., Hinner, M., Marrink, S.J., Tieleman, D.P. A molecular view of cholesterol flip-flop and chemical potential in different membrane environments. J. Am. Chem. Soc. 2009, in press.
49. de Joannis, J., Jiang, Y., Yin, F., Kindt, J.T. Equilibrium distributions of dipalmitoyl phosphatidylcholine and dilauroyl phosphatidylcholine in a mixed lipid bilayer: atomistic semigrand canonical ensemble simulations. J. Phys. Chem. B 2006, 110, 25875–82.
50. Wang, H., de Joannis, J., Jiang, Y., Gaulding, J.C., Albrecht, B., Yin, F., Khanna, K. et al. Bilayer edge and curvature effects on partitioning of lipids by tail length: atomistic simulations. Biophys. J. 2008, 95, 2647–57.
51. Coppock, P.S., Kindt, J.T. Atomistic simulations of mixed-lipid bilayers in gel and fluid phases. Langmuir 2009, 25, 352–9.
52. Rodriguez, Y., Mezei, M., Osman, R. Association free energy of dipalmitoylphosphatidylserines in a mixed dipalmitoylphosphatidylcholine membrane. Biophys. J. 2007, 92, 3071–80.
53. Sapay, N., Bennett, W.F.D., Tieleman, D.P. Thermodynamics of flip-flop and desorption for a systematic series of phosphatidylcholine lipids. Soft Matter 2009, in press.
54. Bennett, W.F.D., MacCallum, J.L., Tieleman, D.P. Thermodynamic analysis of the effect of cholesterol on dipalmitoylphosphatidylcholine lipid membranes. J. Am. Chem. Soc. 2009, 131, 1972–78.
55. Sengupta, D., Leontiadou, H., Mark, A.E., Marrink, S.J. Toroidal pores formed by antimicrobial peptides show significant disorder. Biochim. Biophys. Acta 2008, 1778, 2308–17.
56. Herce, H.D., Garcia, A.E. Molecular dynamics simulations suggest a mechanism for translocation of the HIV-1 TAT peptide across lipid membranes. Proc. Natl. Acad. Sci. U.S.A. 2007, 104, 20805–10.
57. Leontiadou, H., Mark, A.E., Marrink, S.J. Antimicrobial peptides in action. J. Am. Chem. Soc. 2006, 128, 12156–61.

58. Freites, J.A., Tobias, D.J., von Heijne, G., White, S.H. Interface connections of a transmembrane voltage sensor. Proc. Natl. Acad. Sci. U.S.A. 2005, 102, 15059–64.
59. Monticelli, L., Robertson, K.M., MacCallum, J.L., Tieleman, D.P. Computer simulation of the KvAP voltage-gated potassium channel: steered molecular dynamics of the voltage sensor. FEBS Lett. 2004, 564, 325–32.
60. Sands, Z.A., Sansom, M.S. How does a voltage sensor interact with a lipid bilayer? Simulations of a potassium channel domain. Structure 2007, 15, 235–44.
61. Tieleman, D.P. The molecular basis of electroporation. BMC Biochem. 2004, 5, 10.
62. Bockmann, R.A., de Groot, B.L., Kakorin, S., Neumann, E., Grubmuller, H. Kinetics, statistics, and energetics of lipid membrane electroporation studied by molecular dynamics simulations. Biophys. J. 2008, 95, 1837–50.
63. Tarek, M. Membrane electroporation: a molecular dynamics simulation. Biophys. J. 2005, 88, 4045–53.
64. Vernier, P.T., Ziegler, M.J. Nanosecond field alignment of head group and water dipoles in electroporating phospholipid bilayers. J. Phys. Chem. B 2007, 111, 12993–6.
65. Kandasamy, S.K., Larson, R.G. Cation and anion transport through hydrophilic pores in lipid bilayers. J. Chem. Phys. 2006, 125, 074901.
66. Gurtovenko, A.A., Vattulainen, I. Pore formation coupled to ion transport through lipid membranes as induced by transmembrane ionic charge imbalance: atomistic molecular dynamics study. J. Am. Chem. Soc. 2005, 127, 17570–1.
67. Gurtovenko, A.A., Vattulainen, I. Ion leakage through transient water pores in protein-free lipid membranes driven by transmembrane ionic charge imbalance. Biophys. J. 2007, 92, 1878–90.
68. Vernier, P.T., Ziegler, M.J., Sun, Y., Gundersen, M.A., Tieleman, D.P. Nanopore-facilitated, voltage-driven phosphatidylserine translocation in lipid bilayers–in cells and in silico. Phys. Biol. 2006, 3, 233–47.
69. Notman, R., den Otter, W.K., Noro, M.G., Briels, W.J., Anwar, J. The permeability enhancing mechanism of DMSO in ceramide bilayers simulated by molecular dynamics. Biophys. J. 2007, 93, 2056–68.
70. Gurtovenko, A.A., Onike, O.I., Anwar, J. Chemically induced phospholipid translocation across biological membranes. Langmuir 2008, 24, 9656–60.
71. Gurtovenko, A.A., Anwar, J. Ion transport through chemically induced pores in protein-free phospholipid membranes. J. Phys. Chem. B 2007, 111, 13379–82.
72. Tieleman, D.P., Leontiadou, H., Mark, A.E., Marrink, S.J. Simulation of pore formation in lipid bilayers by mechanical stress and electric fields. J. Am. Chem. Soc. 2003, 125, 6382–3.
73. Leontiadou, H., Mark, A.E., Marrink, S.J. Molecular dynamics simulations of hydrophilic pores in lipid bilayers. Biophys. J. 2004, 86, 2156–64.
74. MacCallum, J.L., Tieleman, D.P. Interactions between small molecules and lipid bilayers. Comput. Mod. Mem. Bilayers 2008, 60, 227–56.
75. MacCallum, J.L., Bennett, W.F.D., Tieleman, D.P. Distribution of amino acids in a lipid bilayer from computer simulations. Biophys. J. 2008, 94, 3393–404.
76. Yoo, J., Cui, Q. Does arginine remain protonated at the lipid membrane? Insights from microscopic pKa calculations. Biophys. J. 2008, 94, L61–3.
77. Li, L., Vorobyov, I., MacKerell, A.D. Jr., Allen, T.W. Is arginine charged in a membrane? Biophys. J. 2008, 94, L11–3.
78. Tolpekina, T.V., den Otter, W.K., Briels, W.J. Nucleation free energy of pore formation in an amphiphilic bilayer studied by molecular dynamics simulations. J. Chem. Phys. 2004, 121, 12060–6.
79. Tolpekina, T.V., den Otter, W.K., Briels, W.J. Simulations of stable pores in membranes: system size dependence and line tension. J. Chem. Phys. 2004, 121, 8014–20.
80. Wohlert, J., den Otter, W.K., Edholm, O., Briels, W.J. Free energy of a trans-membrane pore calculated from atomistic molecular dynamics simulations. J. Chem. Phys. 2006, 124, 154905.
81. Notman, R., Anwar, J., Briels, W.J., Noro, M.G., den Otter, W.K. Simulations of skin barrier function: free energies of hydrophobic and hydrophilic transmembrane pores in ceramide bilayers. Biophys. J. 2008, 95, 4763–71.

82. Gurtovenko, A.A., Vattulainen, I. Molecular mechanism for lipid flip-flops. J. Phys. Chem. B 2007, 111, 13554–59.

83. Steck, T.L., Ye, J., Lange, Y. Probing red cell membrane cholesterol movement with cyclodextrin. Biophys. J. 2002, 83, 2118–25.

84. Rog, T., Stimson, L.M., Pasenkiewicz-Gierula, M., Vattulainen, I., Karttunen, M. Replacing the cholesterol hydroxyl group with the ketone group facilitates sterol flip-flop and promotes membrane fluidity. J. Phys. Chem. B 2008, 112, 1946–52.

85. Kucerka, N., Perlmutter, J.D., Pan, J., Tristram-Nagle, S., Katsaras, J., Sachs, J.N. The effect of cholesterol on short- and long-chain monounsaturated lipid bilayers as determined by molecular dynamics simulations and X-ray scattering. Biophys. J. 2008, 95, 2792–805.

CHAPTER **2**

Quantifying Uncertainty and Sampling Quality in Biomolecular Simulations

Alan Grossfield[1] and **Daniel M. Zuckerman**[2]

Contents

[1] Department of Biochemistry and Biophysics, University of Rochester Medical Center, Rochester, NY, USA
[2] Department of Computational Biology, University of Pittsburgh School of Medicine, Pittsburgh, PA, USA

Annual Reports in Computational Chemistry, Volume 5
ISSN: 1574-1400, DOI 10.1016/S1574-1400(09)00502-7

23

Abstract   Growing computing capacity and algorithmic advances have facilitated the study of increasingly large biomolecular systems at longer timescales. However, with these larger, more complex systems come questions about the quality of sampling and statistical convergence. What size systems can be sampled fully? If a system is not fully sampled, can certain "fast variables" be considered well converged? How can one determine the statistical significance of observed results? The present review describes statistical tools and the underlying physical ideas necessary to address these questions. Basic definitions and ready-to-use analyses are provided, along with explicit recommendations. Such statistical analyses are of paramount importance in establishing the reliability of simulation data in any given study.

Keywords: error analysis; principal component; block averaging; convergence; sampling quality; equilibrium ensemble; correlation time; ergodicity

1. INTRODUCTION

It is a well-accepted truism that the results of a simulation are only as good as the statistical quality of the sampling. To compensate for the well-known sampling limitations of conventional molecular dynamics (MD) simulations of even moderate-size biomolecules, the field is now witnessing the rapid proliferation of multiprocessor computing, new algorithms, and simplified models. These changes underscore the pressing need for unambiguous measures of sampling quality. Are current MD simulations long enough to make quantitative predictions? How much better are the new algorithms than the old? Can even simplified models be fully sampled?

Overall, errors in molecular simulation arise from two factors: inaccuracy in the models and insufficient sampling. The former is related to choices in representing the system, for example, all-atom vs. coarse grained models, fixed charge vs. polarizable force fields, and implicit vs. explicit solvent, as well as technical details like the system size, thermodynamic ensemble, and integration algorithm used. Taken in total, these choices define the model used to represent the system of interest. The second issue, quality of sampling, is largely orthogonal to the choice of model. In some sense, assessing the quality of the sampling is a way of asking how accurately a given quantity was computed *for the chosen model*. While this review will focus on the issue of sampling, it is important to point out that without adequate sampling, the predictions of the force fields remain unknown: very few conclusions, positive or negative, can be drawn from an undersampled calculation. Those predictions are embodied most directly in the equilibrium ensemble that simulations have apparently failed to produce in all but small-molecule systems [1,2]. Thus, advances in force field design and parameterization for large biomolecules must proceed in parallel with sampling advances and their assured quantification.

This review will attempt to acquaint the reader with the most important ideas in assessing sampling quality. We will address both the statistical uncertainty in individual observables and quantification of the global quality of the equilibrium ensemble. We will explicitly address differing approaches necessary for standard dynamics simulations, as compared to algorithms such as replica exchange, and while we use the language of MD, virtually all of the arguments apply equally to Monte Carlo (MC) methods as well. Although this review will not specifically address path sampling, many of the ideas carry over to what amounts to equilibrium sampling of the much larger space of paths. We will recommend specific "best practices," with the inevitable bias toward the authors' work. We have tried to describe the intellectual history behind the key ideas, but the article is ultimately organized around practically important concepts.

For the convenience of less experienced readers, key terms and functions have been defined in the appendix: average, variance, correlation function, and correlation time.

1.1 Examples big and small: butane and rhodopsin

Example trajectories from small and large systems (to which we will return throughout the review) illustrate the key ideas. In fact, almost all the complexity we will see in large systems is already present in a molecule as simple as *n*-butane. Nevertheless, it is very valuable to look at both "very long" trajectories and some that are "not long enough." Concerning the definition of "long," we hope that if "we know it when we see it," then we can construct a suitable mathematical definition. Visual confirmation of good sampling is still an important check on any quantitative measure.

1.1.1 Butane

Let us first consider butane, as in Figure 1. Several standard molecular coordinates are plotted for a period of 1 ns, and it is clear that several timescales less than 1 ns are present. The very fastest motions (almost vertical in the scale of the figure) correspond to bond length and angle vibrations, while the dihedrals exhibit occasional quasi-discrete transitions. The CH_3 dihedral, which reports on methyl spinning, clearly makes more frequent transitions than the main dihedral.

Perhaps the trajectory of butane's C–C–C angle is most ambiguous, since there appears to be a slow overall undulation in addition to the rapid vibrations. The undulation appears to have a frequency quite similar to the transition rate of the main dihedral, and underscores the point that *generally speaking, all degrees of freedom are coupled*, as sketched in Figure 2. In the case of butane, the sampling quality of the C–C–C angle may indeed be governed by the slowest motions of the molecule and isomerization of the central torsion.

1.1.2 Rhodopsin

It is perhaps not surprising that all of the degrees of freedom are tightly coupled in a simple system like butane. It seems reasonable that this coupling may be less important in larger biomolecular systems, where there are motions on timescales

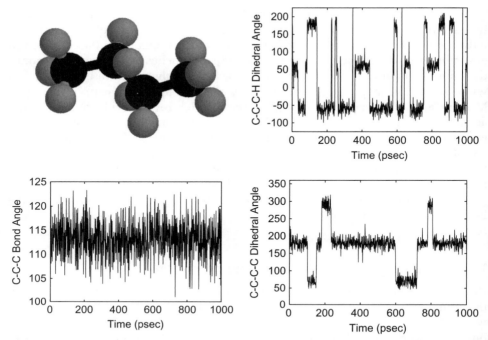

Figure 1 Widely varying timescales in *n*-butane. Even the simple butane molecule (upper left) exhibits a wide variety of dynamical timescales, as exhibited in the three time traces. Even in the fast motions of the C–C–C bond angle, a slow undulation can be detected visually.

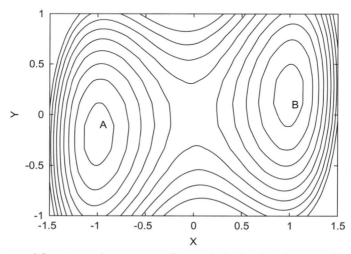

Figure 2 Slow and fast timescales are generally coupled. The plot shows a schematic two-state potential. The *y* coordinate is fast regardless of whether state A or B is occupied. However, fast oscillations of *y* are no guarantee of convergence because the motions in *x* will be much slower. In a molecule, all atoms interact — even if weakly or indirectly — and such coupling must be expected.

ranging from femtoseconds to milliseconds; indeed, it is commonly assumed that small-scale reorganizations, such as side-chain torsions in proteins, can be computed with confidence from MD simulations of moderate length. While this assumption is likely true in many cases, divining the cases when it holds can be extremely difficult. As a concrete example, consider the conformation of the retinal ligand in dark-state rhodopsin. The ligand is covalently bound inside the protein via Schiff base linkage to an internal lysine, and contains an aromatic hydrocarbon chain terminated by an ionone ring. This ring packs against a highly conserved tryptophan residue, and is critical to this ligand's role as an inverse agonist.

The ring's orientation, relative to the hydrocarbon chain, is largely described by a single torsion, and one might expect that this kind of local quantity would be relatively easy to sample in a MD simulation. The quality of sampling for this torsion would also seem easy to assess, because as for most torsions, there are three stable states. However, Figure 3 shows that this is not the case, because of coupling between fast and slow modes. The upper frame of Figure 3 shows a time series of this torsion from a MD simulation of dark-state rhodopsin [3]; the three expected torsional states ($g+$, $g-$, and t) are all populated, and there are a number of transitions, so most practitioners would have no hesitation in concluding that (a) the trajectory is reasonably well sampled, and (b) that all three states are frequently populated, with $g-$ the most likely and *trans* the least. The middle panel, however, shows the same trajectory extended to 150 ns; it too seems to suggest a clear conclusion, in this case that the transitions in the first 50 ns are part of a slow equilibration, but that once the protein has relaxed the retinal is stable in the $g-$ state. The bottom panel, containing the results of extending the trajectory to 1,600 ns, suggests yet another distinct conclusion, that $g-$ and t are the predominant states, and rapidly exchange with each other, on the nanosecond scale.

These results highlight the difficulties involved in assessing the convergence of single observables. No amount of visual examination of the upper and middle panels would have revealed the insufficiency of the sampling (although it is interesting to note that the "effective sample size" described below is not too large). Rather, it is only after the fact, in light of the full 1,600 ns trajectory, that the sampling flaws in the shorter trajectories become obvious. This highlights the importance of considering timescales broadly when designing and interpreting simulations. This retinal torsion is a local degree of freedom, and as such should relax relatively quickly, but the populations of its states are coupled to the conformation of the protein as a whole. As a result, converging the sampling for the retinal requires reasonable sampling of the protein's internal degrees of freedom, and is thus a far more difficult task than it would first appear.

1.2 Absolute vs. relative convergence

Is it possible to describe a simulation as absolutely converged? From a statistical point of view, we believe the answer is clearly "no," except in those cases where

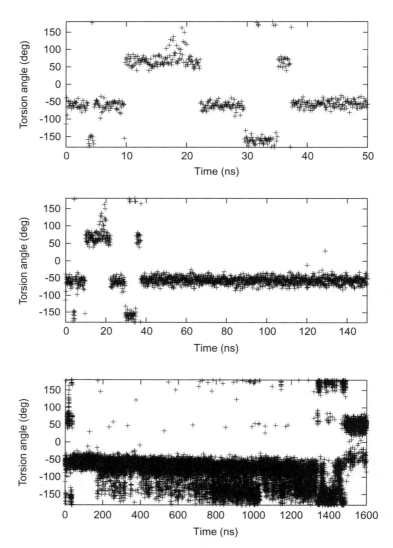

Figure 3 Time series for the torsion connecting the ionone ring to the chain of rhodopsin's retinal ligand. All three panels show the same trajectory, cut at 50, 150, and 1,600 ns, respectively.

the correct answer is already known by other means. Whether a simulation employs ordinary MD or a much more sophisticated algorithm, so long as the algorithm correctly yields canonical sampling according to the Boltzmann factor, one can expect the statistical quality will increase with the duration of the simulation. In general, the statistical uncertainty of most conceivable molecular simulation algorithms will decay inversely with the square root of simulation length. The square-root law should apply once a stochastic simulation process is in the true sampling regime — that is, once it is long enough to produce multiple properly distributed statistically independent configurations.

The fundamental perspective of this review is that simulation results are not absolute, but rather are intrinsically accompanied by statistical uncertainty [4–8]. Although this view is not novel, it is at odds with informal statements that a simulation is "converged." Beyond quantification of uncertainty for specific observables, we also advocate quantification of overall sampling quality in terms of the "effective sample size" [8] of an equilibrium ensemble [9,10].

As a conceptual rule-of-thumb, any estimate for the average of an observable which is found to be based on fewer than ~ 20 statistically independent configurations (or trajectory segments) should be considered unreliable. There are two related reasons for this. First, any estimate of the uncertainty in the average based on a small number of observations will be unreliable because this uncertainty is based on the variance, which converges more slowly than the observable (i.e., average) itself. Second, any time the estimated number of statistically independent observations (i.e., effective sample size) is ~ 20 or less, both the overall sampling quality and the sample-size estimate itself must be considered suspect. This is again because sample-size estimation is based on statistical fluctuations that are, by definition, poorly sampled with so few independent observations.

Details and "best practices" regarding these concepts will be given below.

1.3 Known unknowns

1.3.1 Lack of ergodicity — unvisited regions of configuration space
No method known to the authors can report on a simulation's failure to visit an important region of configuration space unless these regions are already known in advance. Thus, we instead focus on assessing sampling quality in the regions of space that has been visited. One can hope that the generation of many effectively independent samples in the known regions of configuration space with a correct algorithm is good "insurance" against having missed parts of the space — but certainly it is no guarantee. Larger systems are likely to have more thermodynamically relevant substates, and may thus require more independent samples even in the absence of significant energetic barriers.

1.3.2 Small states rarely visited in dynamical simulation
This issue is also related to ergodicity, and is best understood through an example. Consider a potential like that sketched in Figure 4, with two states of 98% and 2% population at the temperature of interest. A "perfect" simulation capable of generating fully independent configurations according to the associated Boltzmann factor would simply yield 2 of every 100 configurations in the small state, on average. However, a finite dynamical simulation behaves differently. As the barrier between the states gets larger, the frequency of visiting the small state will decrease exponentially. Thus, estimating an average like $\langle x \rangle$ will be very difficult — since the small state might contribute appreciably. Further, quantifying the uncertainty could be extremely difficult if there are only

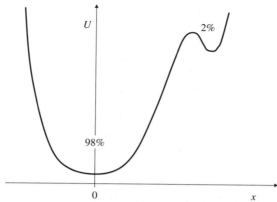

Figure 4 Cartoon of a landscape for which dynamical simulation is intrinsically difficult to analyze. As the barrier between the states gets higher, the small state requires exponentially more dynamical sampling, even though the population may be inconsequential. It would seem that, in principle, a cutoff should be chosen to eliminate "unimportant" states from analysis. In any complex molecular system, there will always be extremely minor but almost inaccessible basins.

a small number of visits to the small state — because the variance will be poorly estimated.

1.4 Non-traditional simulation methods

The preceding discussion applied implicitly to what we classify as dynamical simulations — namely, those simulations in which all correlations in the final trajectory arise because each configuration is somehow generated from the previous one. This time-correlated picture applies to a broad class of algorithms: MD, Langevin and Brownian dynamics, as well as traditional Monte Carlo (MC, also known as Markov-chain Monte Carlo). Even though MC may not lead to true physical dynamics, all the correlations are sequential.

However, in other types of molecular simulation, any sampled configuration may be correlated with configurations *not* sequential in the ultimate "trajectory" produced. That is, the final result of some simulation algorithms is really *a list of configurations, with unknown correlations,* and not a true trajectory in the sense of a time series.

One increasingly popular method which lead to non-dynamical trajectories is replica-exchange MC or MD [11–13], which employs parallel simulations at a ladder of temperatures. The "trajectory" at any given temperature includes repeated visits from a number of (physically continuous) trajectories wandering in temperature space. Because the continuous trajectories are correlated in the usual sequential way, their intermittent — that is, non-sequential — visits to the various specific temperatures produce *non*-sequential correlations when one of those temperatures is considered as a separate ensemble or "trajectory" [14]. Less prominent examples of non-dynamical simulations occur in a broad class of polymer-growth algorithms (e.g., refs. 15–17).

Because of the rather perverse correlations that occur in non-dynamical methods, there are special challenges in analyzing statistical uncertainties and sampling quality. This issue has not been well explored in the literature; see however [10,18,19]. We therefore present some tentative thoughts on non-dynamical methods, based primarily on the notion that independent simulations appear to provide the most definitive means for analyzing non-dynamical simulations. In the case of replica exchange, understanding the difference between "mixing" and sampling will prove critical to any analysis.

2. ERROR ESTIMATION IN SINGLE OBSERVABLES

One of the main goals of biomolecular simulation is the estimation of ensemble averages, which should always be qualified by estimates of statistical uncertainty. We will review the two main approaches to estimating uncertainty in averages, but a general note of caution should be repeated. Because all variables can be correlated in a complex system, the so-called "fast" variables may not be as fast as they appear based on standard error estimation techniques: see Figure 2. As in the examples of the rhodopsin dihedral, above, even a single coordinate undergoing several transitions may not be well sampled. Also, investigators should be wary of judging overall sampling quality based on a small number of observables unless they are specifically designed to measure ensemble quality, as discussed below.

The present discussion will consider an arbitrary observable f, which is a function of the configuration $\mathbf{x}$ of the system being simulated. The function $f(\mathbf{x})$ could represent a complex measure of an entire macromolecule, such as the radius of gyration, or it could be as simple as a single dihedral or distance.

Our focus will be on time correlations and block averaging. The correlation-time analysis has been in use for some decades [7], including to analyze the first protein MD simulation [20], and it embodies the essence of all the single-observable analyses known to the authors. The block-averaging approach [5,21] is explicitly described below because of its relative simplicity and directness in estimating error using simple variance calculations. Block averaging "short cuts" the need to calculate a correlation time explicitly, although timescales can be inferred from the results. Similarly, the "ergodic measure" of Thirumalai and coworkers [22–24], not described here, uses variances and correlation times implicitly.

Both the correlation-time analysis and the block-averaging scheme described below assume that a *dynamical trajectory* is being analyzed. Again, by "dynamical" we only mean that correlations are "transmitted" via sequential configurations — which is not true in a method like replica exchange.

2.1 Correlation-time analysis

The correlation-time analysis of a single observable has a very intuitive underpinning. Consider first that dynamical simulations (e.g., molecular and

Langevin dynamics), as well as "quasi-dynamical" simulations (e.g., typical MC [25]), create trajectories that are correlated solely based on the sequence of the configurations. As described in the Appendix, the correlation time τ_f measures the length of simulation time — whether for physical dynamics or MC — required for the trajectory to lose "memory" of earlier values of f. Therefore, the correlation time τ_f *for the specific observable f* provides a basis for estimating the number of statistically independent values of f present in a simulation of length t_{sim}, namely $N_f^{ind} \sim t_{sim}/\tau_f$. By itself, $N_f^{ind} \gg 1$ would suggest good sampling for the particular observable f.

The correlation time is computed from the correlation function (see Appendix), and it is useful to consider an example. Figure 5 shows the time-correlation functions computed for individual state lifetimes as measured by a 100 ns simulation of butane. Specifically, for each snapshot from the trajectory, the central torsion was classified as *trans*, $g+$, or $g-$. A time series was then written for each state, with a value of 1 if the system was in that state and 0 otherwise. The autocorrelation functions for each of those time series are shown in Figure 5. All three correlation functions drop smoothly to zero within 200 ps, suggesting that a 100 ns simulation should contain a very large number of independent samples. However, the populations for the three states over the course of the trajectory are 0.78, 0.10, and 0.13 for the *trans*, $g+$, and $g-$ states, respectively. The $g+$ and $g-$ states are physically identical, and thus should have the same populations in the limit of perfect sampling. Thus even a very long simulation of a very simple system is incapable of estimating populations with high precision.

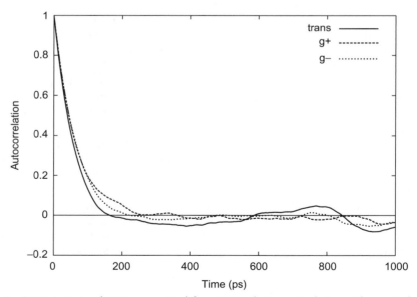

Figure 5 State autocorrelations computed from 100 ns butane simulations. The central torsion was labeled as either *trans*, $g+$, or $g-$, and the autocorrelation function for presence in each state was computed.

To obtain an estimate of the statistical uncertainty in an average $\langle f \rangle$, the correlation time τ_f must be used in conjunction with the variance σ_f^2 (square of the standard deviation; see Appendix) of the observable. By itself, the standard deviation only gives the basic scale or range of fluctuations in f, which might be much larger than the uncertainty of the *average* $\langle f \rangle$. In other words, it is possible to know very precisely the average of a quantity that fluctuates a lot: as an extreme example, imagine measuring the average height of buildings in Manhattan. In a dynamical trajectory, the correlation time τ_f provides the link between the range of fluctuations and the precision (uncertainty) in an average, which is quantified by the standard error of the mean, SE,

$$\mathrm{SE}(f) = \frac{\sigma_f}{\sqrt{N_f^{\mathrm{ind}}}} \sim \sigma_f \sqrt{\frac{\tau_f}{t_{\mathrm{sim}}}} \tag{1}$$

In this notation, N_f^{ind} is the number of independent samples contained in the trajectory, and t_{sim} the length of the trajectory. The standard error can be used to approximate confidence intervals, with a rule of thumb being that $\pm 2\mathrm{SE}$ represents *roughly* a 95% confidence interval [26]. The actual interval depends on the underlying distribution and the sampling quality as embodied in $N_f^{\mathrm{ind}} \sim t_{\mathrm{sim}}/\tau_f$; see ref. 25 for a more careful discussion.

It has been observed that the simple relation between correlation time and sampling quality embodied in the estimate $N_f^{\mathrm{ind}} = t_{\mathrm{sim}}/\tau_f$ is actually *too conservative* in typical cases [27]. That is, even though the simulation may require a time τ_f to "forget" its past (with respect to the observable f), additional information beyond a single estimate for f is obtained in the period of a single correlation time — that is, from partially correlated configurations. However, the improvement in sampling quality is modest — the effective sample size may be double the estimate based simply on τ_f. Such subtleties are accounted for automatically in the block-averaging analysis described below.

Understanding the correlation-time analysis, as well as the habitual calculation of correlation functions and times, is extremely useful. Yet the analysis has weaknesses for quantifying uncertainty that suggest relying on other approaches for generating publication-quality error bars. First, like any single-observable analysis, the estimation of correlation times may fail to account for slow timescales in observables not considered: recall the rhodopsin example. Second, the calculation of correlation times becomes less reliable in precisely those situations of greatest interest — when a second, slower timescale enters the intrinsically noisier tail of the correlation function. The third weakness was already described: a lack of full accounting for all statistical information in the trajectory. These latter considerations suggest that a block-averaging procedure, described next, is a preferable analysis of a single observable.

2.2 Block averaging

When executed properly, the block-averaging analysis automatically corrects two of the weaknesses in correlation-time estimates of the error based on equation (1). In particular, any slow timescales present in the time series for the particular observable are accounted for (although *only* in regard to the observable studied). Second, because block averaging uses the full trajectory, it naturally includes all the information present. The block-averaging analysis was first reported by Flyvbjerg and Petersen [5], who credited the previously unpublished idea to others.

The approach can be described simply (although it is not easily understood from the original reference). A trajectory with $N = M \times n$ snapshots is divided into M segments ("blocks"), with an initial very short block length, such as $n = 1$ (see Figure 6). The average of the observable is calculated for each block yielding M values for $\langle f \rangle_i$, with $i = 1,\ldots,M$. The block length n is gradually increased and the set of block averages is recalculated for each length. Further, for each value of n, the standard deviation among the block averages, σ_n, is used to calculate a running estimate of the overall standard error, namely,

$$\text{BSE}(f, n) = \frac{\sigma_n}{\sqrt{M}} \tag{2}$$

This is the standard error in estimates of the mean based on blocks (trajectory segments) of length n. Clearly, for small n (and large $M = N/n$) when consecutive blocks are highly correlated, blocked standard error (BSE) greatly underestimates the statistical error, since equation (2) only yields the true standard error when all M blocks are statistically independent. On the other hand, once the blocks are essentially independent of one another (i.e., when the block length is substantially greater than the correlation time, $n \gg \tau_f/\Delta t$), BSE will cease to vary with n and become a reliable estimator of the true SE. Figure 6 illustrates this behavior for a trigonometric function of butane's main (C–C–C–C) dihedral.

The function $\text{BSE}(f, n)$ therefore increases monotonically with n and asymptotes to the true standard error associated with $\langle f \rangle$, as seen in Figure 6. Thus, a plot of $\text{BSE}(f, n)$ includes a "signal" as to whether or not the error estimate has converged, which is not subject to the extremes of numerical uncertainty associated with the tail of a correlation function. Furthermore, the block-averaging analysis directly includes all trajectory information (all frames).

The only weakness of the block-averaging approach, which is minor in our opinion, is that it does not directly render the correlation time. Having the correlation time in hand provides important physical intuition. Nevertheless, we note that the correlation time can be estimated cleanly using the block-averaging results. Specifically, using the trajectory $f(t)$, one can directly calculate the variance σ_f and then solve for N_f^{ind} using Equation (1). The correlation time is then given approximately by $\tau_f \sim t_{\text{sim}}/N_f^{\text{ind}}$, which will somewhat underestimate the correlation time (as noted implicitly by Berg and Harris, ref. 27) perhaps by a factor of ~ 2.

It is not uncommon for researchers to use the name "block averaging" to describe a second, far simpler procedure. In this case, a single time series is split into M blocks, and the variance between the averages for those blocks is presented as the uncertainty. However, unlike the true block-averaging protocol described above, this procedure is not statistically meaningful, because the

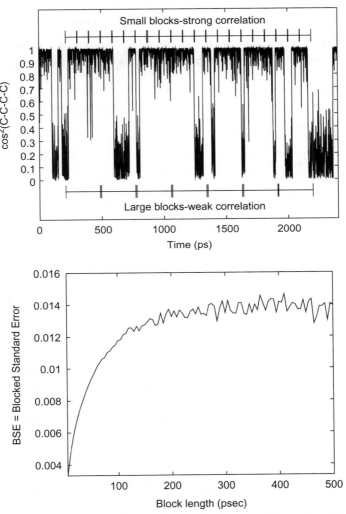

Figure 6   The block-averaging procedure considers a full range of block sizes. The upper panel shows the time series for the squared cosine of the central dihedral of butane, with two different block sizes annotated. The lower panel shows the block-averaged standard error for that times series, as a function of block size.

single-block size is chosen arbitrarily; it is only by systematically varying the block size that one can reliably draw conclusions about the uncertainty.

2.3 Summary — single observables in dynamical simulations

Several points are worth emphasizing: (i) single observables should not be used to assess overall sampling quality. (ii) The central ideas in single-observable analysis are that the correlation time separates statistically independent values of the observable, and that one would like to have many statistically independent "measurements" of the observable — that is, $N_f^{\text{ind}} \gg 1$. (iii) The block-averaging analysis is simple to implement and provides direct estimation of statistical uncertainty. We recommend that the correlation time and effective sample size also be estimated to ensure $N_f^{\text{ind}} \gg 1$. (iv) In a correlation-time analysis, one wants to ensure the total simulation time is a large multiple of the correlation time — that is, $t_{\text{sim}}/\tau_f \gg 1$.

2.4 Analyzing single observables in non-dynamical simulations

As discussed earlier, the essential fact about data from non-dynamical simulations (e.g., replica exchange and polymer-growth methods) is that a configuration occurring at one point in the "trajectory" may be highly correlated with another configuration anywhere else in the final list of configurations. Similarly, a configuration could be fully independent of the immediately preceding or subsequent configurations. To put it most simply, the list of configurations produced by such methods is not a time series, and so analyses based on the explicit or implicit notion of a correlation time (time correlations are implicit in block averaging) cannot be used.

From this point of view, the only truly valid analysis of statistical errors can be obtained by considering independent simulations. Ideally, such simulations would be started from different initial conditions to reveal "trapping" (failure to explore important configurational regions) more readily. Running multiple simulations appears burdensome, but it is better than excusing "advanced" algorithms from appropriate scrutiny. Of course, rather than multiplying the investment in computer time, the available computational resources can be divided into 10 or 20 parts. All these parts, after all, are combined in the final estimates of observable averages. Running independent trajectories is an example of an "embarrassingly parallel" procedure, which is often the most efficient use of a standard computer cluster. Moreover, if a simulation method is not exploring configuration space well in a tenth of the total run time, then it probably is not performing good sampling anyway.

How can statistical error be estimated for a single observable from independent simulations? There seems little choice but to calculate the standard error in the mean values estimated from each simulation using Equation (1), where the variance is computed among the averages from the independent simulations and N_f^{ind} is set to the number of simulations. In essence, each simulation is treated as a single measurement, and presumed to be totally independent of the other trajectories. Importantly, one can perform a "reality check" on such a calculation because the variance of the observable can also be calculated from all data from all simulations — rather than from the simulation means. The squared ratio of this absolute variance to the variance of the means yields a separate (albeit crude) estimate of the number of independent samples. This latter estimate should be of the same order as, or greater than, the number of independent simulations, indicating that each "independent" simulation indeed contained at least one statistically independent sample of the observable.

It is interesting to observe that, in replica-exchange simulations, the physically continuous trajectories (which wander in temperature) can be analyzed based on time-correlation principles [10,14]. Although each samples a non-traditional ensemble, it is statistically well defined and can be used as a proxy for the regular ensemble. A more careful analysis could consider, separately, those segments of each continuous trajectory at the temperature of interest. The standard error among these estimates could be compared to the true variance, as above, to estimate sampling quality. A detailed discussion of these issues in the context of weighted histogram analysis of replica-exchange simulation is given by Chodera et al. [14].

3. OVERALL SAMPLING QUALITY IN SIMULATIONS

In contrast to measures of convergence that reflect a single local observable, for example, a torsion angle, some methods focus on the global sampling quality. For a simulation of a macromolecule, the distinction would be between asking "how well do I know this particular quantity?" and "how well have I explored the conformational space of the molecule?" The latter question is critical, in that if the conformational space is well sampled, most physical quantities should be known well.

This review will describe two classes of analyses of overall sampling quality: (i) qualitative and visual techniques, which are mainly useful in convincing oneself a simulation is *not* sufficiently sampled; and (ii) quantitative analyses of sampling, which estimate the "effective sample size."

3.1 Qualitative and visual analyses of overall sampling effectiveness

There are a number of techniques that, although they cannot quantitatively assess convergence or statistical uncertainty, can give tremendous qualitative insight. While they cannot tell the user that the simulation has run long enough, they can quickly suggest that the simulation has *not* run long enough. Thus, while they should not replace more rigorous methods like block averaging and sample-size estimation, they are quite useful.

3.1.1 Scalar RMSD analyses

One of the simplest methods is the comparison of the initial structure of the macromolecule to that throughout the trajectory via a distance measure such as the root mean square deviation (RMSD). This method is most informative for a system like a folded protein under native conditions, where the molecule is expected to spend the vast majority of the time in conformations quite similar to the crystal structure. If one computes the RMSD time series against the crystal structure, one expects to see a rapid rise due to thermal fluctuations, followed by a long plateau or fluctuations about a mean at longer timescales. If the RMSD time series does not reach a steady state, the simulation is either (a) still equilibrating or (b) drifting away from the starting structure. In any event, until the system assumes a steady-state value — one that may fluctuate significantly, but has no significant trend — the system is clearly not converged. Indeed, one can argue that under that circumstance equilibrium sampling has not yet even begun. However, beyond this simple assessment, RMSD is of limited utility, mostly because it contains little information about what states are being sampled; a given RMSD value maps a 3N-dimensional hypersphere of conformation space (for N atoms) to a single scalar, and for all but the smallest RMSD values this hypersphere contains a broad range of structures. Moreover, the limiting value for the RMSD cannot be known in advance. We know the value should be non-zero and not large, but the expected plateau value is specific to the system studied, and will vary not only between macromolecules, but also with changes to simulation conditions such as temperature and solvent.

An improvement is to use a windowed RMSD function as a measure of the rate of conformation change. Specifically, for a given window length (e.g., 10 consecutive trajectory snapshots), the average of the all of the pairwise RMSDs (or alternatively, the average deviation from the average over that interval) is computed as a function of time. This yields a measure of conformational diversity over time, and can more readily reveal conformational transitions.

3.1.2 All-to-all RMSD analysis

A more powerful technique is to compute the RMSDs for all pairs of snapshots from the trajectory and plot them on a single graph [28]. Figure 7 shows the results of such a plot, made using the alpha-carbon RMSD computed from a 1.6 μs all-atom simulation of dark-state rhodopsin in an explicit lipid membrane [3]. The plot reveals a hierarchical block structure along the diagonal; this suggests that the protein typically samples within a substate for a few hundred nanoseconds, and then rapidly transitions to a new conformational well. However, with the exception of two brief excursions occurring around 280 and 1,150 ns into the trajectory, the system never appears to leave and then return to a given substate. This suggests that this simulation, although very long by current standards, probably has not fully converged.

3.1.3 Cluster counting

A more general approach, also based on pairwise distances, would be to use cluster analysis. Although a general discussion of the many clustering algorithms presently in use is beyond the scope of this manuscript, for our purposes we define clustering to be any algorithm that divides an ensemble into sets of self-similar structures. One application of clustering to the assessment of convergence came from Daura et al., who measured the rate of discovery of new clusters over the course of a trajectory; when this rate became very small, the simulation was presumed to be reasonably well converged [29]. However, a closer look reveals this criterion to be necessary but not sufficient to guarantee good sampling. While it is true that a simulation that is still exploring new states is unlikely to have achieved good statistics (at least for a reasonable definition of "states"),

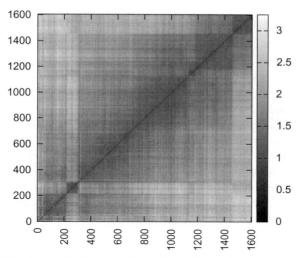

Figure 7 All-to-all RMSD for rhodopsin alpha-carbons. The scale bar to the right shows darker grays to indicate a more similar structures.

simply having visited most of the thermodynamically relevant states is no guarantee that a simulation will produce accurate estimates of observables.

3.1.4 "Structural histogram" of clusters

As discussed by Lyman and Zuckerman [9], not only must clusters be visited, but also it is important that the populations of those regions be accurately reproduced, since the latter provide the weights used to compute thermo-dynamic averages. In a procedure building on this idea, one begins by performing a cluster analysis on the entire trajectory to generate a vocabulary of clusters or bins. The cluster/bin populations can be arrayed as a one-dimensional "structural histogram" reflecting the full configuration-space distribution. Structural histograms from parts of the trajectory are compared to one computed for the full trajectory, and plotting on a log-scale gives the variation in knot units, indicating the degree of convergence.

3.1.5 Principal components analysis

Principal component analysis (PCA) is another tool that has been used extensively to analyze molecular simulations. The technique, which attempts to extract the large-scale characteristic motions from a structural ensemble, was first applied to biomolecular simulations by Garcia [28], although an analogous technique was used by Levy et al. [30]. The first step is the construction of the $3N \times 3N$ (for an N-atom system) fluctuation correlation matrix

$$C_{ij} = \langle x_i - \bar{x}_i \rangle \langle x_j - \bar{x}_j \rangle$$

where x_i represents a specific degree of freedom (e.g., the z-coordinate of the 23rd atom) and the overbar indicates the average structure. This task is commonly simplified by using a subset of the atoms from the molecule of interest (e.g., the α-carbons from the protein backbone). The matrix is then diagonalized to produce the eigenvalues and eigenvectors; the eigenvectors represent character-istic motions for the system, while each eigenvalue is the mean square fluctuation along its corresponding vector. The system fluctuations can then be projected onto the eigenvectors, giving a new time series in this alternative basis set. The diagonalization and time series projection can be performed efficiently using singular value decomposition, first applied to principal component analysis of biomolecular fluctuations by Romo et al. [31].

In biomolecular systems characterized by fluctuations around a single structure (e.g., equilibrium dynamics of a folded protein), a small number of modes frequently account for the vast majority of the motion. As a result, the system's motions can be readily visualized, albeit abstractly, by plotting the time series for the projections of the first two or three modes. For example, projecting the rhodopsin trajectory described above [3] onto its two largest principle modes yields Figure 8. As with the all-to-all RMSD plots (see Figure 7), this method readily reveals existence of a number of substates, although temporal information is obscured. A well-sampled simulation would exhibit a large number of transitions among substates, and the absence of significant transitions can readily be visualized by plotting principal components against time. It is important to note that this method does not depend on the physical significance or statistical convergence of the eigenvectors themselves, which is reassuring because previous work has shown that these vectors can be extremely slow to converge [1,32]. Rather, for these purposes the modes serve as a convenient coordinate system for viewing the motions.

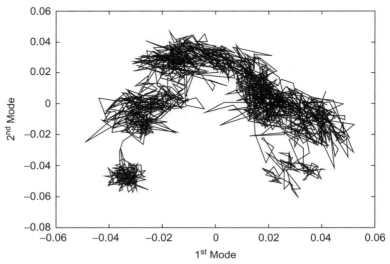

Figure 8   Projection of rhodopsin fluctuations onto the first two modes derived from principal component analysis. As with Figure 7, this method directly visualizes substates in the trajectory.

PCA can also be used to quantify the degree of similarity in the fluctuations of two trajectories (or two portions of a single trajectory). The most rigorous measure is the covariance overlap suggested by Hess [1,33,34]

$$\Omega_{A:B} = 1 - \left[\frac{\sum_{i=1}^{3N}(\lambda_i^A + \lambda_i^B) - 2\sum_{i=1}^{3N}\sum_{j=1}^{3N}\sqrt{\lambda_i^A \lambda_j^B}(\vec{v}_i^A \cdot \vec{v}_j^B)}{\sum_{i=1}^{3N}(\lambda_i^A + \lambda_i^B)} \right]$$

which compares the eigenvalues λ and eigenvectors v computed from two datasets A and B. The overlap ranges from 0, in the case where the fluctuations are totally dissimilar, to 1, where the fluctuation spaces are identical. Physically, the overlap is in essence the sum of all the squared dot products of all pairs of eigenvectors from the two simulations, weighted by the magnitudes of their displacements (the eigenvalues) and normalized to go from 0 to 1. Hess used this quantity as an internal measure of convergence, comparing the modes computed from subsets of a single trajectory to that computed from the whole [34]. More recently, Grossfield et al. computed the principal components from 26 independent 100 ns simulations of rhodopsin, and used the covariance overlap to quantify the similarity of their fluctuations, concluding that 100 ns is not sufficient to converge the fluctuations of even individual loops [1]. Although these simulations are not truly independent (they used the same starting structure for the protein, albeit with different coordinates for the lipids and water), the results again reinforce the point that the best way to assess convergence is through multiple repetitions of the same system.

3.2 Quantifying overall sampling quality: the effective sample size

To begin to think about the quantification of overall sampling quality — that is, the quality of the equilibrium ensemble — it is useful to consider "ideal sampling" as a reference point. In the ideal case, we can imagine having a perfect computer program which outputs single configurations drawn completely at random and distributed according to the appropriate Boltzmann factor for the system of interest. Each configuration is fully independent of all others generated by this ideal machinery, and is termed "i.i.d." — independent and identically distributed.

Thus, given an ensemble generated by a particular (non-ideal) simulation, possibly consisting of a great many "snapshots," the key *conceptual* question is: To how many i.i.d. configurations is the ensemble equivalent in statistical quality? The answer is the *effective sample size* [1,9,10] which will quantify the statistical uncertainty in every slow observable of interest — and many "fast" observables also, due to coupling, as described earlier.

The key *practical* question is: How can the sample size be quantified? Initial approaches to answering this question were provided by Grossfield et al. [1] and by Lyman and Zuckerman [10]. Grossfield et al. employed a bootstrap analysis to a set of 26 independent trajectories for rhodopsin, extending the previous "structural histogram" cluster analysis [10] into a procedure for estimating sample size. They compared the variance in a cluster's population from the independent simulations to that computed using a bootstrap analysis (bootstrapping is a technique where a number of artificial datasets are generated by choosing points randomly from an existing dataset [35]). Because each data point in the artificial datasets is truly independent, comparison of the bootstrap and observed variances yielded estimates of the number of independent data points (i.e., effective sample size) per trajectory. The results were astonishingly small, with estimates ranging from 2 to 10 independent points, depending on the portion of the protein examined. Some of the numerical uncertainties in the approach may be improved by considering physical states rather than somewhat arbitrary clusters; see below.

Lyman and Zuckerman suggested a related method for estimating sample size [10]. First, they pointed out that binomial and related statistics provided an analytical means for estimating sample size from cluster-population variances, instead of the bootstrap approach. Second, they proposed an alternative analysis *specific to dynamical trajectories*, but which also relied on comparing observed and ideal variances. In particular, by generating observed variances from "frames" in a dynamical trajectory separated by a fixed amount of time, it can be determined whether those time-separated frames are statistically independent. The separation time is gradually increased until ideal statistics are obtained, indicating independence. The authors denoted the minimum time for independence the "structural decorrelation time" to emphasize that the full configuration-space ensemble was analyzed based on the initial clustering/binning.

3.2.1 Looking to the future: can state populations provide a "universal indicator"?

The ultimate goal for sample size assessment (and thus estimation of statistical error) is a "universal" analysis, which could be applied blindly to dynamical or non-dynamical simulations and reveal the effective size. Current work in the Zuckerman group (unpublished) suggests a strong candidate for a universal indicator of sample size is the variance observed from independent simulations in the populations of physical states. Physical states are to be distinguished from the more arbitrary clusters discussed above, in that a state is characterized by relatively fast timescales internally, but slow timescales for transitions between states. (Note that proximity by RMSD or similar distances does not indicate either of these properties.) There are two reasons to focus on populations of physical states: (i) the state populations arguably are *the* fundamental description of the equilibrium ensemble, especially considering that (ii) as explained below, relative state populations cannot be accurate unless detailed sampling *within states* is correct. Of course, determining physical states is non-trivial but apparently surmountable [36].

We claim that if you know state populations, you have sampled well — at least in an equilibrium sense. Put another way, we believe it is impossible to devise an algorithm — dynamical or non-dynamical — that could correctly sample state populations without sampling correctly within states. The reason is that the ratio of populations of any pair of states depends on the ensembles internal to the states. This ratio is governed/defined by the ratio of partition functions for the states, i and j, which encompass the non-overlapping configuration-space volumes V_i and V_j, namely,

$$\frac{\text{prob}(i)}{\text{prob}(j)} = \frac{Z_i}{Z_j} = \frac{\int_{V_i} d\mathbf{r}\ e^{-U(\mathbf{r})/k_B T}}{\int_{V_j} d\mathbf{r}\ e^{-U(\mathbf{r})/k_B T}} \tag{3}$$

This ratio cannot be estimated without sampling within both states — or effectively doing so [37,38]. Note that this argument does not assume that sampling is performed dynamically.

If indeed the basic goal of equilibrium sampling is to estimate state populations, then these populations can act as the fundamental observables amenable to the types of analyses already described. In practical terms, following 10, a binomial description of any given state permits the effective sample size to be estimated from the populations of the state recorded in independent simulations — or from effectively independent segments of a sufficiently long trajectory. This approach will be described shortly in a publication.

One algorithm for blindly approximating physical states has already been proposed [36], although the method requires the number of states to be input. In work to be reported soon, Zhang and Zuckerman developed a simple procedure for approximating physical states that does not require input of the number of states. In several systems, moreover, it was found that sample-size estimation is relatively insensitive to the precise state definitions (providing they are reasonably physical, in terms of the timescale discussion above). The authors are therefore optimistic that a "benchmark" blind, automated method for sample-size characterization will be available before long.

3.3 Analyzing non-standard simulations — for example, replica exchange

The essential intuition regarding non-standard/non-dynamical simulations such as replica exchange has been given in our discussion of single observables: in brief, a given configuration in a "trajectory" may be highly correlated with much "later" configurations, yet not correlated with intervening configurations. Therefore, a reliable analysis must be based on multiple independent simulations — which is perhaps less burdensome than it first seems, as discussed above.

We believe such simulations should be analyzed using state-population variances. This approach, after all, is insensitive to the origins of the analyzed "trajectories" and any internal time correlations or lack thereof. No method that relies explicitly or implicitly on time correlations would be appropriate.

Replica-exchange simulations, because of their growing popularity, merit special attention. While their efficacy has been questioned recently [19,39], our purpose here is solely to describe appropriate analyses. To this end, a clear distinction must be drawn between "mixing" (accepted exchanges) and true

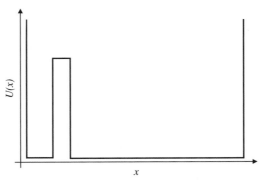

Figure 9 A cartoon of two states differing only in entropy. Generally, in any simulation, energetic effects are much easier to handle than entropic. The text describes the challenge of analyzing errors in replica-exchange simulations when only entropy distinguishes two energetically equal states.

sampling. While mixing is necessary for replica exchange to be more efficient than standard dynamics (otherwise each temperature is independent), *mixing in no way suggests good sampling has been performed*. This can be clearly appreciated from a simple "thought experiment" of a two-temperature replica-exchange simulation of the double square well potential of Figure 9. Assume the two replicas have been initiated from different states. Because the states are exactly equal in energy, every exchange will be accepted. Yet if the barrier between the states is high enough, no transitions will occur in either of the physically continuous trajectories. In such a scenario, replica exchange will artifactually predict 50% occupancy of each state. A block averaging or time-correlation analysis of a single temperature will not diagnose the problem. As suggested in the single-observable discussion, some information on under-sampling may be gleaned from examining the physically continuous trajectories. The most reliable information, however, will be obtained by comparing multiple independent simulations; Section 2.4 explains why this is cost efficient.

4. RECOMMENDATIONS

1. *General.* When possible, perform multiple simulations, making the starting conformations as independent as possible. This is recommended regardless of the sampling technique used.
2. *Single observables.* Block averaging is a simple, relatively robust procedure for estimating statistical uncertainty. Visual and correlation analyses should also be performed.
3. *Overall sampling quality — heuristic analysis.* If the system of interest can be thought of as fluctuating about one primary structure (e.g., a native protein), use qualitative tools, such as projections onto a small number of PCA modes or all-to-all RMSD plots to simplify visualization of trajectory quality. Such

heuristic analyses can readily identify under-sampling as a small number of transitions.

4. *Overall sampling quality — quantitative analysis.* For dynamical trajectories, the "structural decorrelation time" analysis [10] can estimate the slowest timescale affecting significant configuration-space populations and hence yield the effective sample size. For non-dynamical simulations, a variance analysis based on multiple runs is called for [1]. Analyzing the variance in populations of approximate physical states appears to be promising as a benchmark metric.

5. *General.* No amount of analysis can rescue an insufficiently sampled simulation. A smaller system or simplified model that has been sampled well may be more valuable than large detailed model with poor statistics.

ACKNOWLEDGMENTS

D.M. Zuckerman would like to acknowledge in-depth conversations with Edward Lyman and Xin Zhang, as well as their assistance in preparing the figures. Insightful discussions were also held with Divesh Bhatt, Ying Ding, Artem Mamonov, and Bin Zhang. Support for DMZ was provided by the NIH (Grants GM076569 and GM070987) and the NSF (Grant MCB-0643456). AG would like to thank Tod Romo for helpful conversations and assistance in figure preparation.

REFERENCES

1. Grossfield, A., Feller, S.E., Pitman, M.C. Convergence of molecular dynamics simulations of membrane proteins. Proteins Struct. Funct. Bioinformatics 2007, 67, 31–40.
2. Shirts, M.R., Pande, V.S. Solvation free energies of amino acid side chain analogs for common molecular mechanics water models. J. Chem. Phys. 2005, 122, 144107.
3. Grossfield, A., Pitman, M.C., Feller, S.E., Soubias, O., Gawrisch, K. Internal hydration increases during activation of the G-protein-coupled receptor rhodopsin. J. Mol. Biol. 2008, 381, 478–86.
4. Binder, K., Heermann, D.W. Monte Carlo Simulation in Statistical Physics: An Introduction, 2nd edn., Springer, Berlin, 1988.
5. Flyvbjerg, H., Petersen, H.G. Error estimates on averages of correlated data. J. Chem. Phys. 1989, 91, 461–6.
6. Ferrenberg, A.M., Landau, D.P., Binder, K. Statistical and systematic errors in Monte Carlo sampling. J. Stat. Phys. 1991, 63, 867–82.
7. Binder, K., Heermann, D.W. Monte Carlo Simulation in Statistical Physics: An Introduction, 3rd edn., Springer, Berlin, 1997.
8. Janke, W. In Quantum Simulations of Complex Many-Body Systems: From Theory to Algorithms (eds J. Grotendorst, D. Marx and A. Muramatsu), Vol. 10, John von Neumann Institute for Computing, Julich, 2002, pp. 423– 45.
9. Lyman, E., Zuckerman, D.M. Ensemble based convergence assessment of biomolecular trajectories. Biophys. J. 2006, 91, 164–72.
10. Lyman, E., Zuckerman, D.M. On the structural convergence of biomolecular simulations by determination of the effective sample size. J. Phys. Chem. B 2007, 111, 12876–82.
11. Swendsen, R.H., Wang, J.-S. Replica Monte Carlo simulation of spin-glasses. Phys. Rev. Lett. 1986, 57, 2607.
12. Geyer, C. J. Proceedings of the 23rd Symposium on the Interface Interface Foundation, 1991.
13. Sugita, Y., Okamoto, Y. Replica-exchange molecular dynamics method for protein folding. Chem. Phys. Lett. 1999, 314, 141–51.

14. Chodera, J.D., Swope, W.C., Pitera, J.W., Seok, C., Dill, K.A. Use of the weighted histogram analysis method for the analysis of simulated and parallel tempering simulations. J. Chem. Theory Comput. 2007, 3, 26–41.
15. Wall, F.T., Erpenbeck, J.J. New method for the statistical computation of polymer dimensions. J. Chem. Phys. 1959, 30, 634–7.
16. Grassberger, P. Pruned-enriched Rosenbluth method: Simulations of theta polymers of chain length up to 1 000 000. Phys. Rev. E 1997, 56, 3682–93.
17. Liu, J.S. Monte Carlo Strategies in Scientific Computing, Springer, New York, 2002.
18. Denschlag, R., Lingenheil, M., Tavan, P. Efficiency reduction and pseudo-convergence in replica exchange sampling of peptide folding–unfolding equilibria. Chem. Phys. Lett. 2008, 458, 244–8.
19. Nymeyer, H. How efficient is replica exchange molecular dynamics? An analytic approach. J. Chem. Theory Comput. 2008, 4, 626–36.
20. McCammon, J.A., Gelin, B.R., Karplus, M. Dynamics of folded proteins. Nature 1977, 267, 585–90.
21. Kent, D.R., Muller, R.P., Anderson, A.G., Goddard, W.A., Feldmann, M.T. Efficient algorithm for "on-the-fly" error analysis of local or distributed serially correlated data. J. Comput. Chem. 2007, 28, 2309–16.
22. Mountain, R.D., Thirumalai, D. Measures of effective ergodic convergence in liquids. J. Phys. Chem. 1989, 93, 6975–9.
23. Thirumalai, D., Mountain, R.D. Ergodic convergence properties of supercooled liquids and glasses. Phys. Rev. A 1990, 42, 4574.
24. Mountain, R.D., Thirumalai, D. Quantative measure of efficiency of Monte Carlo simulations. Physica A 1994, 210, 453–60.
25. Berg, B.A. Markov Chain Monte Carlo Simulations and Their Statistical Analysis, World Scientific, New Jersey, 2004.
26. Spiegel, M.R., Schiller, J., Srinivasan, R.A. Schaum's Outline of Probability and Statistics, 2nd edn, McGraw Hill, New York, 2000.
27. Berg, B.A., Harris, R.C. From data to probability densities without histograms. Comput. Phys. Commun. 2008, 179, 443–8.
28. Garcia, A.E. Large-amplitude nonlinear motions in proteins. Phys. Rev. Lett. 1992, 68, 2696–9.
29. Smith, L.J., Daura, X., Gunsteren, W.F.v. Assessing equilibration and convergence in biomolecular simulations. Proteins 2002, 48, 487–96.
30. Levy, R.M., Srinivasan, A.R., Olson, W.K., Mccammon, J.A. Quasi-harmonic method for studying very low-frequency modes in proteins. Biopolymers 1984, 23, 1099–112.
31. Romo, T.D., Clarage, J.B., Sorensen, D.C., Phillips, G.N. Automatic identification of discrete substates in proteins-singular-value decomposition analysis of time-averaged crystallographic refinements. Proteins 1995, 22, 311–21.
32. Balsera, M.A., Wriggers, W., Oono, Y., Schulten, K. Principal component analysis and long time protein dynamics. J. Phys. Chem. 1996, 100, 2567–72.
33. Faraldo-Gomez, J.D., Forrest, L.R., Baaden, M., Bond, P.J., Domene, C., Patargias, G., Cuthbertson, J., Sansom, M.S.P. Conformational sampling and dynamics of membrane proteins from 10-nanosecond computer simulations. Proteins Struct. Funct. Bioinformatics 2004, 57, 783–91.
34. Hess, B. Convergence of sampling in protein simulations. Phys. Rev. E 2002, 65, 031910.
35. Efron, B., Tibshirani, R.J. An Introduction to the Bootstrap, Chapman and Hall, CRC, Boca Raton, FL, 1998.
36. Chodera, J.D., Singhal, N., Pande, V.S., Dill, K.A., Swope, W.C. Automatic discovery of metastable states for the construction of Markov models of macromolecular conformational dynamics. J. Chem. Phys. 2007, 126, 155101–17.
37. Voter, A.F. A Monte Carlo method for determining free-energy differences and transition state theory rate constants. J. Chem. Phys. 1985, 82, 1890–9.
38. Ytreberg, F.M., Zuckerman, D.M. Peptide conformational equilibria computed via a single-stage shifting protocol. J. Phys. Chem. B 2005, 109, 9096–103.
39. Zuckerman, D.M., Lyman, E. A second look at canonical sampling of biomolecules using replica exchange simulation. J. Chem. Theory Comput. 2006, 2, 1200–2.
40. Boon, J.P., Yip, S. Molecular Hydrodynamics, Dover, New York, 1992.

APPENDIX

For reference, we provide brief definitions and discussions of basic statistical quantities: the mean, variance, autocorrelation function, and autocorrelation time.

Mean

The mean is simply the average of a distribution, which accounts for the relative probabilities of different values. If a simulation produces a correct distribution of values of the observable f, then relative probabilities are accounted for in the set of N values sampled. Thus the mean $\langle f \rangle$ is estimated via

$$\langle f \rangle = \frac{1}{N} \sum_{i=1}^{N} f_i \tag{A.1}$$

where f_i is the ith value recorded in the simulation.

Variance

The variance of a quantity f, which is variously denoted by σ_f^2, var(f), or $\sigma^2(f)$, measures the intrinsic range of fluctuations in a system. Given N properly distributed samples of f, the variance is defined as the average squared deviation from the mean:

$$\sigma_f^2 = \langle (f - \langle f \rangle)^2 \rangle = \frac{1}{N-1} \sum_{i=1}^{N} (f_i - \langle f \rangle)^2 \tag{A.2}$$

The factor of $N-1$ in the denominator reflects that the mean is computed from the samples, rather than supplied externally, and one degree of freedom is effectively removed.

The square root of the variance, the standard deviation, σ_f, thus quantifies the width or spread in the distribution; it has the same units as f itself, unlike the variance. Except in specialized analyses (such a block averaging) the variance does not quantify error. As an example, the heights of college students can have a broad range — that is, large variance — while the average height can be known with an error much smaller than the standard deviation.

Autocorrelation function

The autocorrelation function quantifies, on a unit scale, the degree to which a quantity is correlated with values of the same quantity at later times. The function can be meaningfully calculated for any dynamical simulation, in the sense defined earlier, and therefore including MC. We must consider a set of time-ordered values of the observable of interest, so that $f_j = f(t = j\Delta t)$, with $j = 1, 2, \ldots, N$ and Δt the time step between frames. (For MC simulations, one can simply set $\Delta t \equiv 1$). The average amount of autocorrelation between "snapshots" separated by a time t' is quantified by

$$
\begin{aligned}
c_f(t') &= \frac{\langle [f(t) - \langle f \rangle][f(t + t') - \langle f \rangle] \rangle}{\sigma_f^2} \\
&= \frac{(1/N)\sum_{j=1}^{N-(t'/\Delta t)}[f(j\Delta t) - \langle f \rangle]\ [f(j\Delta t + t') - \langle f \rangle]}{\sigma_f^2}
\end{aligned}
\tag{A.3}
$$

where the sum must prevent the argument of the second f from extending beyond N. Note that for $t' = 0$, the numerator is equal to the variance, and the correlation is maximal at the value $c_f(0) = 1$. As t' increases significantly, for any given j, the later values of f are as likely to be above the mean as below it — independent of f_i since the later values have no "memory" of the earlier value. Thus, the correlation function begins at one and decays to zero for long enough times. It is possible for c_f to become negative at intermediate times — which suggests a kind of oscillation of the values of f.

Correlation time

The (auto)correlation time τ_f quantifies the amount of time necessary for simulated (or even experimental) values of f to lose their "memory" of earlier values. In terms of the autocorrelation function, we can say roughly that the correlation time is smallest t' value for which $c_f(t') \ll 1$ for all subsequent times (within noise). More quantitatively, the correlation time can be defined via

$$
\tau_f = \int_0^\infty dt'\ c_j(t')
\tag{A.4}
$$

where the numerical integration must be handled carefully due to the noise in the long-time tail of the correlation function. More approximately, the correlation time can be fit to a presumed functional form, such as an exponential or a sum of exponentials, although it is not necessarily easy to predetermine the appropriate form [40].

CHAPTER **3**

Methods for Monte Carlo Simulations of Biomacromolecules

Andreas Vitalis and **Rohit V. Pappu**

Contents

Abstract

The state-of-the-art for Monte Carlo (MC) simulations of biomacromolecules is reviewed. Available methodologies for sampling conformational equilibria and associations of biomacromolecules in the canonical ensemble, given a continuum description of the solvent environment, are reviewed. Detailed sections are provided dealing with the choice of degrees of freedom, the efficiencies of MC algorithms and algorithmic peculiarities, as well as the optimization of simple movesets. The issue of introducing correlations into elementary MC moves and the applicability of such methods to simulations of biomacromolecules are discussed. A brief discussion of multicanonical methods and an overview of recent simulation work highlighting the potential of MC methods are also provided. It is argued that MC simulations, although underutilized in the biomacromolecular simulation community, hold promise for simulations of complex

Department of Biomedical Engineering, Molecular Biophysics Program, Center for Computational Biology, Washington University in St. Louis, St. Louis, MO, USA

Annual Reports in Computational Chemistry, Volume 5
ISSN: 1574-1400, DOI 10.1016/S1574-1400(09)00503-9

49

systems and phenomena that span multiple length scales, especially when used in conjunction with implicit solvation models or other coarse-graining strategies.

Keywords: Monte Carlo simulations; polypeptides; polynucleotides; concerted rotations; multicanonical ensemble; torsional space

1. INTRODUCTION

Unlike in other fields of computational science, Monte Carlo (MC) sampling is not used very often for simulating biomacromolecules. Historically, this can be explained by a strong bias toward all-atom models of biomacromolecular systems including both macromolecule(s) and solvent. All physically realistic Hamiltonians employ excluded volume and other stiff, nonbonded interactions leading to very rugged energy landscapes and large correlations between many degrees of freedom in these dense systems [1,2]. For such systems, gradient-based techniques such as molecular dynamics (MD [3]) are assumed to be vastly superior, although this notion has been challenged [4]. With the advent of refined implicit solvent models [5–7] and the resultant smoother energy landscapes, MC simulations may gain in appeal for the simulation community. Here, we provide an overview of the state-of-the-art MC methods designed for efficient sampling of biomacromolecules using implicit solvent models in an effective isothermal–isochoric (NVT) ensemble. The use of MC for simulating chains in other ensembles or for simulations of all-atom condensed-phase systems is only touched upon briefly. For more general purposes, the textbook by Frenkel and Smit [8] remains an excellent resource as do other overviews in similar textbooks [9,10].

The Markov chain Metropolis scheme [11] is by far the most common MC methodology. The system is randomly perturbed and the proposed move from microstate A to B is accepted with probability:

$$p_{A\to B} = \max\left[1, \exp(-\beta \cdot \Delta E)\right] \tag{1}$$

Here, ΔE is the energy difference to transition from state A to B and β the reciprocal thermal energy. Metropolis et al. [11] showed that such a scheme samples the Boltzmann distribution associated with the given Hamiltonian at the temperature specified by β. For larger systems, such importance sampling is vastly superior to any systematic or random enumeration schemes, which scale extremely poorly with the number of degrees of freedom in the system [8].

In principle, the transition from state A to B needs to be unbiased and ergodic [12,13]. An illustrative example comes from n-butane, where the degree of freedom for MC sampling is the single backbone dihedral angle, ϕ. Transitions between microstates are made by randomly choosing values for ϕ from an interval. If some values of ϕ are chosen with higher probability, then the proposed transitions are sampled from a biased — as opposed to an unbiased — distribution. Conversely, the sampling could be unbiased, but the values for ϕ

may be sampled from a small subinterval of possibilities. In this case, the simulation would suffer from broken ergodicity because it cannot — by fiat — sample all possible conformations for n-butane. There is a hidden ergodicity issue that pertains to the choices one makes for the degrees of freedom in an MC simulation, a point that is addressed in Section 2.1.

Given a choice for the degrees of freedom, the MC movesets dictate how transitions between different microstates are realized. For biomacromolecules, the movesets have to be diverse. This is important because broken ergodicity can also result from movesets that fail to connect distinct points in conformational space. In any MC simulation, the frequencies with which different elementary moves are attempted *and* the parameters associated with the different move types (such as the maximal displacements associated with different moves) are adjustable. If the movesets are unbiased and ergodic, a sufficiently long MC simulation should ultimately converge to the same result. In practice, computational resources are finite and the choices made for adjustable parameters play an important role in determining whether a simulation actually yields a converged result. Metrics to be used to guide choices for these parameters are discussed in Section 2.3. Two important guidelines are as follows. First, as many choices as possible should be made randomly rather than with a predetermined "schedule." For example, the choice of degree(s) of freedom to perturb during an elementary move should be random rather than scanning all degrees of freedom systematically. Second, it is always true that every problem has its own optimal parameter set. Hence, simulators should always have the freedom to adjust all open parameters.

There are several other reasons for the scarce use of MC in simulations of biomacromolecules. One is the absence of suitable software. The commercial software, BOSS/MCPRO, is provided by the Jorgensen group [14]. The most common, freely available simulations packages tailored to the biomacromolecular simulation community, that is, GROMACS [15], NAMD [16], and TINKER [17], have no MC capability. AMBER [18], which may be the most widely used molecular simulations software, does not provide MC support. However, the CHARMM [19] package now includes an MC module [20]. Some freely available MC programs like MCCCS Towhee [21] are not specifically tailored to biomacromolecular simulations. Others, like PROFASI [22], currently support only very limited Hamiltonians. As detailed in Section 2.2, the software layout for MD and MC codes is fundamentally different, a reason that contributes to the lack of available programs. We hope that our freely available CAMPARI software package, which is scheduled for release in the summer of 2009, will provide a useful addition to the small group of suitable programs such as CHARMM, MCPRO, and PROFASI.

Although MC is an underutilized tool in the field of computational molecular biophysics, there are beneficial features that can be exploited, especially in conjunction with implicit solvent models. Specifically, MC has the potential of accessing length scales that are inaccessible to MD in complex phenomena like peptide aggregation or conformational sampling of intrinsically disordered proteins.

The rest of this review article is structured as follows:

The first issue we discuss is the choice of degrees of freedom. We briefly review the literature pertaining to justification and implementation of performing molecular simulations in non-Cartesian space, most prominently torsional and rigid-body space. We lay out advantages of such a procedure and comment upon common implementation difficulties, with specific attention to the issue of consistency between the development and application of force field parameters for use in MC simulations.

The second area we review is that of computational efficiency of energy evaluations including special considerations for cutoffs and long-range treatment of electrostatic interactions. We provide a brief overview of the required bookkeeping given a Hamiltonian of a certain complexity. Many of these issues are unique to MC calculations due to the fundamentally different ways in which systems evolve when compared to MD calculations.

Next, we discuss strategies to make maximal use of simple unbiased movesets for conformational sampling of biomacromolecules. We provide an example to illustrate how the inclusion of conformational fluctuations spanning multiple length scales can improve the quality of sampling. We conclude with a discussion of the metrics and guidelines that can be used to optimize a moveset for a given system.

The fourth area we cover is that of introducing correlations into MC movesets. We discuss typical biased movesets employed in MC simulations of biomacromolecules and the corrections and parameter settings needed for the incorporation of biased moves. We address concerns pertaining to computational efficiency and ease of implementation.

The fifth section provides a brief description of multicanonical techniques and their use and applicability in MC simulations. We touch upon the benefits that MC techniques provide in sampling ensembles other than the canonical ensemble for biomacromolecular systems.

We conclude by providing a few examples illustrating the peculiarities of sampling phase space via MC for nontrivial systems relevant to the biomacromolecular field. We provide an outlook regarding current challenges and the potential strategies that can be developed or adopted to overcome these challenges.

2. CONFORMATIONAL SAMPLING OF BIOMACROMOLECULES IN THE CANONICAL ENSEMBLE VIA MONTE CARLO METHODS

2.1 Choosing degrees of freedom in conjunction with the force field

One of the benefits of MC algorithms is the ability to naturally deal with constraints, that is, to set up a simulation in terms of arbitrary sets of degrees of freedom, which may well be different from the degrees of freedom over which the potential energy is evaluated. In MD, such functionality is introduced by holonomic constraints [23], for which a variety of popular algorithms have been introduced. A long-stranding issue concerned the introduction of mass-metric tensor artifacts that might arise if one were to restrict sampling to torsional degrees of freedom alone. These artifacts are undoubtedly present in MD simulations, because momenta conjugate to the constrained degrees of freedom are explicitly set to zero, thereby introducing a spurious alteration of the volume element of the configurational integral by the determinant of a reduced mass-metric tensor. Hence, the efficiency one might gain through the use of longer time steps is lost through the inefficiency associated with calculating mass-metric tensor determinants and its gradients. However, MC calculations rest on the separation between the momentum and configurational integrals and hence the spurious coupling between conformational degrees of freedom and volume elements does not arise because there are no momenta to be zeroed out [24]. A simple test illustrates this point: Consider an n-alkane; set the potential energy to be zero for all values of the relevant degrees of freedom, and randomly sample the torsional degrees of freedom; as long as the torsions are sampled from an unbiased distribution, it should follow that all values for the degrees of freedom have equal likelihoods of being realized. This test reveals that Fixman-style [25] corrections are not needed in MC simulations for chain molecules — an advantage that does not prevail for MD simulations with holonomic constraints.

Integrands of configurational integrals are proportional to $\exp[-\beta U(\text{DoF})]$, where β denotes the inverse thermal energy parameter and $U(\text{DoF})$ refers to the potential energy that varies as the degrees of freedom (DoF) assume different values. If $U(\text{DoF})$ is set to zero, then the multidimensional integral over phase space volumes provides an estimate of the size of the relevant conformational space and all combinations for the degrees of freedom should have the same probabilities of being accessed. Any accurate MC algorithm has to reproduce the appropriate unbiased distribution when $U(\text{DoF})$ is set to zero. This sanity check is very useful since it identifies the presence of bias and the possibility of broken ergodicity in a straightforward manner. For example, if the DoF were Cartesian coordinates, and one were using periodic boundary conditions, the degrees of freedom should have a uniform distribution in all three dimensions. Torsional degrees of freedom behave similarly. This makes it possible to sample dihedral angles in MC simulations from a uniform random distribution. The same is not true for degrees of freedom where the configurational space volume element depends on the values assumed by the degrees of freedom. Examples of such degrees of freedom are Euclidean distances and Euler angles [10]. In these cases, Jacobian corrections need to be included in either the acceptance or picking probabilities.

The vast majority of MC simulations of biomacromolecules employ torsional space sampling, which is sometimes augmented by sampling of angular degrees of freedom [26] or even the Cartesian coordinates directly [27]. In the presence of stiff harmonic restraints, step sizes for the latter will typically be vanishingly small. Inclusion of such moves is not a matter of choice. Instead, it is determined by the force field used for the calculations. Every molecular mechanics force field, whether it uses an implicit or explicit representation of the solvent, undergoes a calibration process. Rigorously speaking, the applicability of said force field is only guaranteed within the accuracy of calibration if the sampled degrees of freedom are strictly identical between application and calibration. For instance, introducing holonomic constraints on bond lengths and angles in MD simulations using a force field parameterized in the absence of such constraints is incorrect. Rotational barriers in peptides are known to depend strongly on the flexibility of the stiffer modes [28], including the peptide bond itself [29]. While it is argued in general that equilibrium properties are not affected by typical constraints in a statistically significant manner, even the impact on barriers and hence kinetics might be enough to question their introduction into a force field parameterized in their absence.

The consequences for MC simulations of biomacromolecules are obvious: The choice is to either employ a force field designed specifically in the presence of such constraints (e.g., ref. 30) or — using a diverse enough moveset — make sure to sample the appropriate coordinate space directly. The consequences of ignoring this concern are shown in Figure 1. Here, we compare the temperature-dependent, reversible folding/unfolding of an α-helical peptide (Ace-A_5(AAARA)$_3$A-Nme) for the ABSINTH implicit solvation model and Lennard-Jones parameters [30] coupled to OPLS-AA/L partial charge parameters [31]. We compare results from MC in dihedral space (all constraints present) to results obtained using Langevin

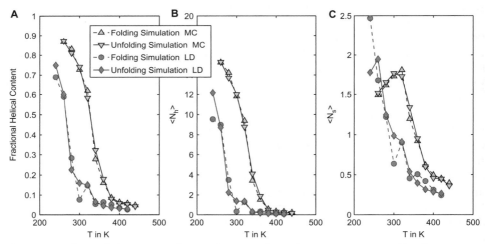

Figure 1   Melting of the FS-peptide as a function of simulation temperature. This figure is analogous to Figure 8 in ref. 30 and the reader is referred there for details of the computation of helix parameters and the description of the MC data. Panel A shows net α-helical content, Panel B the mean number of α-helical hydrogen bonds, and Panel C the mean number of α-helical segments. LD data were obtained using the impulse integrator of Skeel and Izaguirre [32] with uniform friction of $5.0\,ps^{-1}$ and an integration timestep of 1 fs. The total simulation time at each temperature was 25 ns, the first 10 ns of which were discarded for equilibration purposes. No constraints were used, and all bond angle and bond length parameters were adopted directly from OPLS-AA/L [31]. No torsions were employed except those keeping quasi-rigid units planar. Those included the peptide bonds and the guanidinium groups in the arginine side chains.

dynamics (LD [32]) in Cartesian space. The latter used no constraints whatsoever, and bond length and angle parameters were ported from the OPLS-AA/L force field [31]. The MC methodology is identical to what has been published previously [30], while the LD methodology is described in Figure 1. As can be seen, there is a substantial shift in the temperature-midpoint of the melting transition of the α-helix. The additional flexibility decreases helix stability, leading to a lower melting temperature. This result is independent of the metric that we used to quantify the helix–coil transition. It should be noted that some hysteresis error remains for the LD calculation, even though the CPU time needed was larger by a factor of 3–5 than the MC simulation. This efficiency benefit provided by MC sampling is slightly larger than, but generally consistent with, what has been reported in the literature for systems of similar size [26].

In this particular example, constraints strongly affect equilibrium properties as quantified in terms of the response of the system to changes in temperature. This example shows that it is not easy to decouple the effects of constraints on enthalpic barriers from those on minima on the free energy landscape. Caution is therefore required when employing typical biomacromolecular force fields such as AMBER [33], CHARMM [34,35], OPLS [31], or GROMOS [36] in MC simulations in dihedral angle space alone. It is important to ensure that the parameters of force fields have been calibrated using MC as the sampling engine prior to using these parameters in an MC simulation. While this is not the case for simulations of simple systems such as Lennard-Jones fluids or even associative liquids made of rigid molecules, it will certainly be an important consideration for highly flexible polymeric systems.

2.2 Bookkeeping and efficiency in computational algorithms for MC simulations

From a software point of view, it is desirable to have a well-structured hierarchical description for the different biomacromolecules in the simulation system. Such a hierarchy should provide data structures and access functions on the atomic, residue, molecule, and system levels. This allows routines for the evaluation of energy terms to be set up at the level of residue pairs. Experience [14] suggests that this setup is advantageous since it provides a route to easily and intuitively implement the computational algorithms sketched below.

The requirements for data structures and bookkeeping of simulation variables differ fundamentally between MD and MC simulations. In MD programs, forces and energies involving all degrees of freedom are calculated at every step. The system is subsequently evolved and this process is then repeated for the duration of the simulation. The two core assumptions, which are taken advantage of, are as follows: (i) all force and energy evaluations are global, that is, they are performed for the whole system; and (ii) the system evolves in incremental steps such that there is high correlation between the forces and energies from step n to step $n+1$. These assumptions give rise to many algorithmic strategies that are used to speed up MD calculations. These strategies include the use of twin-range cutoffs, neighbor lists that are updated infrequently, and the particle-mesh Ewald (PME [37]) method for treating corrections to long-range interactions.

Unfortunately, neither assumption is true in MC calculations. Instead of requiring global evaluations of forces, evaluations of energy differences ΔE are needed between pairs of microstates that are not necessarily close in phase space, but which might only differ in the values for a subset of degrees of freedom. This implies that a majority of the energy terms remain fixed and they need not be considered when computing ΔE. For the sake of efficiency, this necessitates a strategy for incremental energy updates that is customized for each individual MC move type, which is part of the available moveset. As an example, consider a solution of several small molecules as sketched in Panel A of Figure 2. If we assume a pairwise-additive Hamiltonian and no cutoffs, the terms of the Hamiltonian that require recomputation represent only a small subset of the terms needed to compute the whole-system energy. The computational expense for energy evaluations in this case scales as $O(N)$ with the number of interaction sites N, if the number of molecules is large relative to the number of interaction sites per molecule. For biomacromolecules and their associated movesets, it is important to analyze the efficiency of different elementary move types in these terms. For example, consider the perturbation of an individual backbone dihedral angle in a single biomolecule (Panel B of Figure 2). Depending on the location of

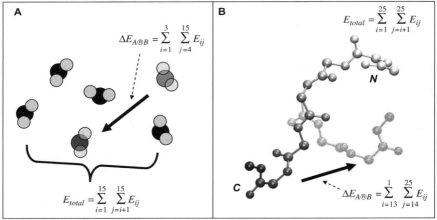

Figure 2 Two examples illustrating complexity of incremental energy updates in MC calculations. Panel A shows the displacement of a single three-site molecule (lighter color) in the presence of four other such molecules. For a strictly pairwise-additive Hamiltonian, the only changing terms are the ones involving the moving molecule (atoms 1–3). Therefore, the resultant energy calculation only needs to encompass 36 of the 105 total pairwise site–site energies. Panel B shows the rotation (pivot) around a single dihedral angle in a molecule composed of 25 sites (assuming systematic numbering from N- to C-terminus). Upon pivoting, the two arms move relative to one another, but remain rigid internally. Of the 300 site–site interactions 156 need to be computed, and the efficiency scales poorly with molecule size (see text).

the residue, the worst complexity for incremental energy evaluations for this type of move is $\sim N^2/2$. For larger molecules, such moves become increasingly less efficient due to the inherent $O(N^2)$ scaling. Conversely, the complexity for a perturbation of a single side-chain degree of freedom is only of order $O(N)$ (analogous to Panel A of Figure 2). Even with cutoffs, the scaling in computational complexity for energy evaluations between different elementary move types will be different. These considerations provide partial motivation for the development and inclusion of truly local MC moves for which scaling ultimately is $O(N)$ (see Section 2.4.1).

Aside from the programming complexity of writing separate wrapper routines for energy updates, the following considerations are peculiar to MC calculations: Not every energy function that is used in simulations of biomacromolecules is suitable for decomposition into static and changing terms. Obviously, a strictly pairwise-additive Hamiltonian is well suited for this type of decomposition. Conversely, non-pairwise-additive Hamiltonians give rise to the complexity that an interaction between two sites is in fact changed as the result of the movement of a third site. This is the case for many implicit solvent models such as the Poisson–Boltzmann (PB [5]) and generalized Born (GB [6]) models, later generations of the EEF1 model [38] and the ABSINTH model [30]. For these models, the effective multibody interactions have to be subjected to a cutoff or similar simplifying assumption to allow efficient use in MC simulations.

In the ABSINTH implicit solvent model, the range of multibody interactions is limited by the size of each atom's (implicit) solvation shell. Since the coupling parameter for these interactions is the solvent-accessible volume, we can keep track of which interactions to recompute in addition to those involving the moving parts during that elementary MC move. We do so by identifying and marking all atoms whose solvent-accessible volumes have changed as a result of the move. All interactions involving at least one of those atoms are then recomputed as well. The strictly local nature of the three-body interactions makes this treatment exact for the ABSINTH model. The scheme also integrates naturally with pairwise cutoffs. Conversely, alternative strategies were proposed for the GB model introducing cutoffs directly into the coupling terms based on their magnitude [39,40]. To correct such an inaccurate treatment in MC simulations, it is important to point to a general strategy cast differently by Gelb [41] and Hetényi et al. [42], which allows the bulk of Metropolis sampling to occur from either a simplified potential or a fast-varying subset of the potential. These short simulation stretches are periodically accepted or rejected using an "outer" Markov chain in order to ensure that the desired distribution is sampled from the full Hamiltonian.

A similar strategy can potentially be used to calculate electrostatic interactions without cutoffs in periodic boundary conditions using the popular PME method [43]. In MC simulations, the same problem arises as outlined in the previous paragraph due to the nondecomposability of the reciprocal space sum. Again, a system-wide evaluation of the full potential energy including the reciprocal part would only occur in regular or random intervals while the bulk of the sampling would be performed from the Hamiltonian given by the real-space sum alone.

An analogous implementation for the standard Ewald method has been presented [44]. Conversely, direct use of the Ewald sum [45] or approximations to it [46–48], which are pairwise decomposable and hence suitable for MC simulations, have generally proven to be too inefficient for most modern applications [49]. Additionally, it should be pointed out that Ewald sums — independent of implementation — are incompatible with implicit solvent models that model a spatially varying dielectric with anything more than trivial functional dependencies [45].

The issue of electrostatics brings up the issue of cutoffs in general. It is beyond the scope of this review to discuss the artifacts introduced by using cutoffs on the electrostatic, excluded volume, and dispersion interactions. It is sufficient to state that for simulations of polar condensed phases or of biomacromolecules in an implicit solvent model such artifacts are hard to detect in the canonical ensemble for system volumes of sufficient size. This stands in contrast to condensed-phase systems with mobile charges (see ref. 50 for an illustration). Regardless, for the MC simulations to remain scalable to larger systems, it is inevitable that some form of simplification of the infinite-range Hamiltonian be considered. A necessary step for distance-dependent potentials is the generation of neighbor lists, that is, the efficient determination of spatial proximity relationships. Here, MC simulations can take advantage of two common strategies:

a. Grid-based methods that operate in $O(N)$ time predominantly known as linked-cell lists [51], which are usually implemented at the level of atoms.
b. Hierarchical methods that take advantage of prior knowledge about molecular topology and that operate in $O(M^2)$ time, where $M = N/n$ and n is the (mean) number of atoms per repeating unit in a polymer, for example, a single protein residue.

With either method, for an energy calculation associated with an elementary move in a large system, a vast majority of the interatomic distances involving the moving parts are never computed. In contrast, Verlet lists [52] are only useful in MC calculations of dense fluids because of the local nature of elementary moves in this particular case.

In summary, strategies developed for MD simulations do not usually apply in MC simulations. Neighbor lists and energies need to be incrementally updated for the trial move and kept or restored to their original values in case of an accepted or rejected move, respectively. Strong emphasis should be placed on computational efficiency. First and foremost, only terms that need to be computed should be computed. As with all softwares, the complexity of algorithms consuming the bulk of CPU time should be carefully analyzed and minimized. Furthermore, strategies using a simplification of the potential and subsequent correction methods or $O(N)$ movesets represent promising avenues for future development. Small enough systems, however, can be dealt with comfortably using even simple movesets. This is the focal point of Section 2.3.

2.3 Optimizing simple MC movesets for biomacromolecular simulations

With the prominent degrees of freedom being dihedral angles (see Section 2.1), a set of straightforward movesets emerges. Polypeptide backbone and side-chain torsions including the peptide bond, but excluding rigid rings or planar systems in the side chains are sampled from a uniform prior distribution. The number of degrees of freedom for each peptide residue therefore ranges from three (Gly) to seven (Lys). For polynucleotides, dihedral angles along the phosphate–sugar backbone are sampled along with rotations around the nucleoside bond as well as any existing carbon–oxygen bonds involving free OH groups [53]. These degrees of freedom can be sampled using simple perturbations of individual or multiple dihedral angles. If such a dihedral angle is part of the polymer backbone, one end of the chain pivots around this joint. Hence, such moves are commonly referred to as pivot moves [54]. In general, moves involving simultaneous perturbations of several degrees of freedom decay exponentially in efficiency with the number of degrees of freedom for a realistic Hamiltonian. This is due to the linearly growing chance of encountering steric clashes, which goes into the energy difference governing the exponential Metropolis criterion (see Equation (1)). Nonetheless, they can be useful as they randomly introduce correlation. The latter may be necessary to convert between conformational states for highly coupled cases like side-chain rotamers [55].

Non-aromatic rings require special treatment. Significant conformational flexibility is retained and a rigid description is inappropriate. The different pucker states of the sugar moiety or the proline side chain are however characterized by a relatively discrete ensemble as is evidenced by analysis of high-resolution crystal structures [56–58]. Simple approaches therefore can be designed and implemented by creating a discrete set of states with an associated energy function meant to reproduce the proper distribution in the context of a suitable background Hamiltonian. Alternatively, specific approaches to find solutions to the ring closure problem may be implemented [59]. Such algorithms are discussed in a different context in Section 2.4.1.

As is well known, the dynamics of polymers in a dense environment under a broad range of conditions become very slow, often glassy [60–63]. A canonical example is that of a single, long polymer in a poor solvent, specifically a mean-field solvent environment in which self-interactions are preferred over chain–solvent interactions [64]. This problem provides a prototype of the complexity encountered in protein-folding problems with implicit solvent models, and hence is of significant relevance to the biomacromolecular field [65]. Aside from complex movesets inspired by such sampling problems (see Section 2.4), what can be done to improve the MC methodology even using simple moves alone? And how does one establish metrics to track sampling efficiency?

We demonstrate the efficiency of a straightforward advancement in simple MC movesets using the peptide Ace-Nle$_{30}$-Nme under "deep quench" conditions. Here, Nle indicates the norleucine residue, which is isosteric with lysine. We monitor the collapse of this peptide from a fully extended state using a degenerate, poor solvent Hamiltonian. Such a Hamiltonian is provided by only employing Lennard-Jones interactions according to parameters published previously [30]. A universal parameter for MC moves is the step size, which is sampled from a uniform distribution over a finite interval. Figure 3 compares three different sampling approaches to pivot moves: (i) all dihedral angles are completely randomized each time they are sampled (maximum step size), (ii) all dihedral angles are perturbed in stepwise fashion, and (iii) both methods are mixed. As can be adjudicated from the relaxation behavior of the system, the

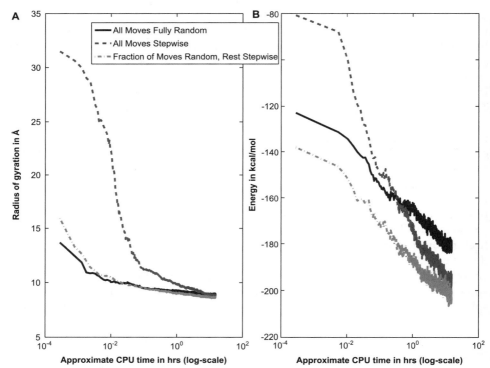

Figure 3 The collapse of the peptide Ace-Nle$_{30}$-Nme under deeply quenched poor solvent conditions monitored by both radius of gyration (Panel A) and energy relaxation (Panel B). MC simulations were performed in dihedral space; 81% of moves attempted to change ϕ/ψ angles, 9% sampled the ω angles, and 10% the side chains. For the randomized case (solid line), all angles were uniformly sampled from the interval $-180°$ to $180°$ each time. For the stepwise case (dashed line), dihedral angles were perturbed uniformly by a maximum of $10°$ for ϕ/ψ moves, $2°$ for ω moves, and $30°$ for side-chain moves. In the mixed case (dash-dotted line), the stepwise protocol was modified to include nonlocal moves with fractions of 20% for ϕ/ψ moves, 10% for ω moves, and 30% for side-chain moves. For each of the three cases, data from 20 independent runs were combined to yield the traces shown. CPU times are approximate, since stochastic variations in runtime were observed for the independent runs. Each run comprised of 3×10^{7} steps. Error estimates are not shown in the interest of clarity, but indicated the results to be robust.

efficiency of approaches (i) and (ii) does seem to track with the density of the system. Full randomizations perform well in the low-density limit, but poorly in the collapse regime. The opposite is true for stepwise perturbations. The important point is that sampling of multiple length scales provides a rigorous, synergistic benefit. The strategy of splitting the moveset into multiple variants of the same basic move type introduces more parameters for the simulator to set. Unfortunately, it is usually not possible to optimize movesets for every problem studied, which implies that intuition and rules of thumb combined with preliminary simulations will inevitably remain prevalent in setting up MC simulations. However, if a more rigorous parameter optimization is needed, we need metrics that can be used to report on the efficiency of sampling.

Aside from relaxation measures such as those shown in Figure 3, metrics of sampling inevitably relate to the rate of "conformational diffusion" or "conformational drift." The first and foremost test in this regard should always be reproducibility by running identical replicas of the same simulation with different starting conditions. Standard deviations of ensemble averages from a sufficient number of independent individual runs yield standard errors, a procedure similar to block averaging, but avoiding all potential correlations between blocks. The magnitude of those standard errors is a high-level but reliable test to guide the optimization of MC movesets. Some systems show reversible order–disorder transitions as a function of control parameters such as temperature. Typical examples of such systems/problems are helix–coil transitions in polypeptides, folding–unfolding transitions of globular proteins, melting transitions of small loop-forming RNA systems, and globule-to-coil transitions in flexible polymers. If disorder-to-order transitions in such systems are reversible, then simulated values of order parameters that track these transitions should yield similar values in the forward and reverse directions. A useful test of a converged MC simulation protocol is to test for reversibility. If there is hysteresis in the simulated order–disorder transitions, then the simulation has clearly not converged. This hysteresis check is directly adopted from experimental work on two-state systems and represents an excellent test within the reduced-dimensional space in which the two-state analysis is performed (see Figure 1). A measure that is similar in spirit is the generalized ergodic measure developed by Straub and Thirumalai [66], which is specifically designed for MD.

Variance- and covariance-based measures such as autocorrelation "times" of instantaneous quantities are useful guides, but are less applicable to MC simulations in particular. First, the absence of significant variance is probably an indicator of a lack of efficient sampling. Second, MC simulations often take discontinuous paths through phase space, which produces substantial stochasticity in such analyses, especially vis-à-vis MD data. Nonetheless, with some prior knowledge of the energy landscape such measures can be used [67]. The bulk of recent work in this area remains dominated by finding efficient ways to cluster simulation data using principal component analysis (PCA) [68], direct clustering techniques [69], or other reduced-dimensional quantities [70,71]. It is then assumed that comprehensive sampling in those spaces is possible and that better coverage and more frequent transitions between basins correspond to improved sampling. Such analyses are typically independent of the sampling engine and therefore suitable for usage in optimizing MC movesets.

Finally, for simulations involving multiple molecules, the rigid-body degrees of freedom have to be sampled. Standard approaches in the spirit of the original Metropolis method are feasible for translational displacement. Rotations of asymmetric particles are conveniently handled by quaternions [72]. One of the cutting-edge applications of MC simulations in the biomacromolecular field is the simulation of peptide aggregation at typical *in vitro*, that is, often very low concentrations. Assuming an implicit solvent representation, the density of such a system is small whereas the volume is large. In MD, this poses the problem of diffusion-limited kinetics, a problem that in certain rare cases may be overcome by adaptive time-step methods [73–75]. The concept of sampling multiple length scales simultaneously applies here in an even more straightforward manner. It is well worth the additional parameters to introduce moves that fully randomize the translational and rotational degrees of freedom of a given molecule. A recent application demonstrates the usefulness of such an approach for the reliable sampling of polyglutamine dimerization at a very low concentration [76].

Of course, a simple moveset will eventually become inefficient with increasing complexity (density and size) of the problem. Hence, diverse attempts have been made to design new and better movesets for MC simulations. Those are discussed next.

2.4 Introducing correlation into MC movesets

As touched upon above, the strategy of introducing correlations into MC simulations by simultaneously changing multiple degrees of freedom is a losing proposition due to increased combinatorial complexity. This is because pivots about multiple torsions lead to rejected moves, especially when the chains are in dense phases. However, correlations are necessary because the simple moves have the complication that the conversions between distinct dense phase configurations are not readily sampled with simple pivot moves. Hence, much effort has been devoted to design movesets that are inherently biased toward inducing a concerted change in the system involving several degrees of freedom at a time. The need for correlation becomes apparent if one considers rugged energy landscapes. More often than not, the paths connecting two adjacent minima will involve a collective degree of freedom, that is, a concerted change in some or all of the relevant elementary degrees of freedom. Such a scenario is sketched in Panel A of Figure 4 for the case of two elementary degrees of freedom. Even if the path is of finite width, elementary moves perturbing only one of the degrees of freedom at a time would be largely ineffective in connecting the two minima. Conversely, an MC move biased toward steering the system along this path would have high acceptance rates despite the need for correcting biases that are introduced by the sampling (see below).

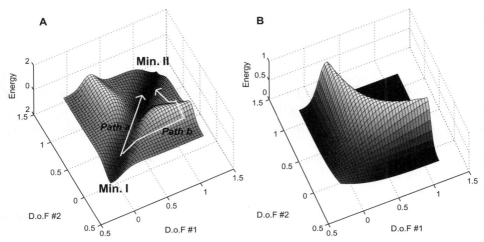

Figure 4 Two arbitrary potential energy surfaces in a two-dimensional coordinate space. All units are arbitrary. Panel A shows two minima connected by a path in phase space requiring correlated change in both degrees of freedom (labeled Path a). As is indicated, paths involving sequential change of the degrees of freedom encounter a large enthalpic barrier (labeled Path b). Panel B shows two minima separated by a barrier. No path with a small enthalpic barrier is available, and correlated, stepwise evolution of the system is not sufficient for barrier crossing.

2.4.1 Concerted rotation (loop-closure) algorithms

A common situation that requires the introduction of correlation into MC movesets is the case in which the ends of the macromolecule itself or of stretches within it are constrained. This is the case for any ring system intended for MC sampling. These include the five-membered rings of sugars and proline, chemically cross-linked macromolecules such as proteins with disulfide linkages, and circular peptides or DNA. However, a much broader range of applications emerges from simply considering a consecutive stretch of residues within a macromolecule. The enclosed stretch or loop is considered on its own. The conformation of the residues in that stretch is resampled while the ends remain in place. These so-called concerted rotation or loop-closure algorithms are attractive for three reasons. First, they are truly local and $O(N)$ and can hence be computationally efficient for a large systems (see Section 2.2). Second, they introduce correlations between multiple degrees of freedom. Third, they lead to local perturbations expected to sample the system much more efficiently than pivot moves at high density, for example, in the interior of a folded protein.

On a lattice, so-called crankshaft moves are trivial implementations of concerted rotations [77]. They have been generalized to the off-lattice case [78] for a simplified protein model. For concerted rotation algorithms that allow conformational changes in the entire stretch, a discrete space of solutions arises when the number of constraints is exactly matched to the available degrees of freedom. The much-cited work by Go and Scheraga [79] formulates the loop-closure problem as a set of algebraic equations for six unknowns reducible effectively to a single equation for a single unknown. The latter is solvable by a systematic search process and eventually yields one or more discrete solutions for all six unknowns. This approach has been recast, modified, and extended multiple times to design algorithms specifically suited to allow local MC moves or exhaustive loop sampling for biopolymers [80–82].

In the context of local MC moves, which are of principal relevance here, we can break the procedure down into several stages:

a. A biased or unbiased presampling or prerotation step perturbing a part of the chain.
b. The chain-closure algorithm solving the constraint problem for six additional degrees of freedom to close the chain exactly.
c. Computation of Jacobian weights for the entire composite move.

Step (a) is usually included in order to ensure that the resultant conformation is significantly different from the previous one. If step (a) were to be omitted, the identity transformation, that is, the solution represented by the conformation the chain is in originally, would always be obtained as a possible solution, which is disadvantageous. The number of degrees of freedom contained in (a) is a free parameter. However, random rotations around multiple dihedrals will easily generate conformations with no solution under the assumption of constant bond lengths. Hence, perturbations around the prerotation segment are typically either small and/or restricted to very few if not a single dihedral angle [83–85], or biased toward keeping the end of the (longer) prerotation segment in roughly the same location [81]. This latter idea was originally introduced by Favrin et al. [86] as an approximate chain-closure technique, a method that, on its own, suffers from the quasi-local nature of the resultant moves.

Step (b) is often the time-consuming one due to the need for performing a systematic search of the solution space in at least one variable. Most implementations suggest a systematic scan of solution space for both forward and reverse moves, which is necessary to properly employ Jacobian-based weighting. Two related advances were proposed: Mezei [85] suggested limiting the search space to conformationally close solutions using a "reverse proximity criterion." It was shown that performance is superior when compared to the more complete version of Hoffmann and Knapp [83]. Similarly, Ulmschneider et al. [81] use a restricted search space with their concerted rotation variant sampling three dihedrals and bond angles each to arrive at a single solution on average. In both cases, the Jacobian needs to be computed only for a single forward and reverse transformation rather than for multiple solutions. This step (c) was ignored in early work and identified and introduced by Dodd et al. [82].

An alternative route to implement local MC moves is provided by the literature on (inverse) kinematics, such as on control systems for robotic arms composed of flexible joints [27,87]. Here, the problem is transformed to either a set of linear equations [27] or finding the roots of a high-order polynomial [87] at comparable computational expense. One of the benefits of such an approach is the ability to introduce arbitrary stiff segments into the loop, that is, the degrees of freedom used for chain closure do not have to be consecutive. Conversely, library-based strategies such as that introduced by Kolodny et al. [88] are not suitable for MC simulations due to the non-ergodicity of the moveset.

In summary, the recommended implementation for concerted rotation moves in MC simulations uses:

i. A prerotation segment of arbitrary length with controllable bias toward keeping the perturbation small
ii. An efficient solution of the closure problem for an arbitrary set of non-consecutive degrees of freedom
iii. Implementations tailored to all biopolymers including polypeptides, polynucleotides, and potentially even lipids as well as polysaccharides

This of course leaves a large set of parameters to optimize for most problems. Only if the simulator has control of all these parameters, can the literature be used to guide those choices. It is beyond the scope of this review to exhaustively test and compare various implementations for MC concerted rotation moves. Naturally, an improvement in the quality of sampling is shown in the original publications for nearly all proposed methods. An interesting data point comes from Ulmschneider et al. [26] who compare MC to MD sampling for a set of small proteins or peptides capable of reversible folding. In a GB/SA implicit solvent model, with the OPLS-AA/L force field, they find that MC sampling, which consists of concerted rotation and simple side-chain moves, is superior by a factor of 2.0–2.5 using folding times as the dominant metric. This compares favorably with our own experience as detailed in Section 2.1 (Figure 1).

2.4.2 Cluster move algorithms

Traditionally, density has been the most crucial limiting factor in making MC techniques useful. At typical liquid densities, acceptance rates even for single-particle moves drop precipitously. Small molecule diffusion is hindered, and the simulation of binary mixtures is near impossible at those densities, and — more importantly — vastly inferior to MD. The reason is of course the extremely high coupling between all degrees of freedom in the system, for which no efficient MC movesets can be designed (see Figure 4).

At slightly lower densities, however, significant advances have been made to introduce correlation into rigid-body moves. The naïve approach is to randomly select two or more molecules and perturb their rigid-body degrees of freedom in concerted fashion, that is, to translate them by the same vector and/or to rotate them around a common point. Such a moveset is perfectly ergodic, unbiased, and potentially able to capture all positive correlation between molecules. Unfortunately, it is highly inefficient and explodes combinatorially with system size. Even for a system of 100 molecules, the chances of picking every possible "cluster" of size 4 are vanishingly small (there are more than 10^8 such unique clusters).

In response, two dominant algorithms were developed initially for on-lattice spin systems [89,90]. Both revolve around determining an effective cluster of spins iteratively based on pairwise energies to construct a move type capable of overcoming the correlation problem in these systems, which were typically observed and referred to as "critical-slowdown phenomena." In constructing the clusters, energetic coupling information is therefore used directly rather than inferring it from spatial coupling. The resultant algorithms can be cast in a rejection-free manner for certain Hamiltonians and differ fundamentally from the Metropolis method. The first algorithm is due to Swendsen and Wang [89], and the second due to Wolff [90]. The literature dealing with methodological advances for both of these methods has been reviewed recently [91]. Off-lattice variants were iteratively refined [92–95] leading to highly efficient movesets for evolving systems of molecules at low enough density, where "low enough" is typically governed by the range and strength of interaction potentials (e.g., see ref. 96). The major drawback of all of these methods is the assumption of a pairwise-additive potential. As outlined in Section 2.2, this is not the case for most modern implicit solvation models that rely on effective many-body interactions. Further development is needed to design an efficient strategy addressing or circumventing this issue. The reader should be reminded, however, that — as an additional complication — none of the above algorithms is known to work for a condensed phase system of strongly interacting particles with long-range interactions. An alternative formalism is proposed by Maggs [97], which might work better in such circumstances.

What is the utility of this body of literature in the context of all-atom simulations of biomacromolecules? Two cutting-edge applications come to mind: First, simulating the aggregation or transient association of peptides and/or proteins with the single-particle Metropolis method is obviously hindered as soon as strongly interacting molecules are present and associate. Due to the low net density, algorithms as presented above represent elegant ways to circumvent those kinetic traps [94]. Second, we anticipate a rapid growth in mixed explicit–implicit solvent models. A simple case is the explicit treatment of electrolyte ions or other cosolutes coupled to a continuum description of the water alone [30]. Often, the cosolutes may represent a dense enough matrix to slow down diffusion of macromolecules even if cosolute molecules do not specifically bind to the latter. Sampling in such mixed-size solutions may be substantially enhanced using appropriate cluster algorithms [96].

2.4.3 Gradient-biased Monte Carlo techniques
One of the oldest ideas in MC simulations is to improve their efficiency by using information about the potential energy gradient. From the outset, this poses two challenges: (i) gradients need to be computed eliminating one of the efficiency benefits of MC over MD, and (ii) for rugged energy landscapes, gradients have only local predictive power; that is, they do not yield guidance for crossing barriers even though the step size is not principally constrained as it is in MD. Dating back to the works of Pangali et al. [98] and Rossky et al. [99], single-particle forces have been used to guide the displacement of particles in dense systems. While a considerable speedup is typically observed relative to the unbiased Metropolis scheme, it remains unclear whether such a method is ultimately superior to dynamics techniques, which ensure explicitly that the canonical ensemble is sampled, that is, stochastic (Langevin) dynamics.

Brownian dynamics (BD), which is stochastic dynamics in the over-damped limit, can just as well be understood as force-biased (dynamic) MC employing collective moves only [100,101].

Not surprisingly, a large body of work has emerged trying to link the methodologies while taking full advantage of the correlation introduced by using gradients. Such correlation is maximal for Newtonian MD due to the absence of random noise, which would cause friction and loss of correlations in momenta (governed by the fluctuation–dissipation theorem). Originally, Duane et al. [102] suggested augmenting MD with Metropolis MC moves to accept only configurations consistent with the canonical ensemble. The requirement in the MD portion is that the integrator be time-reversible and symplectic. This mixing of NVE MD sampling with an outer Markov chain has enabled taking larger time steps in the MD portion compared to straight NVE or NVT MD. The efficiency, however, has been criticized due to rapidly decaying acceptance rates essentially caused by integrator error [103]. Hence, multiple improvements have been suggested [104–108], which extend beyond the scope of this review. An interesting question can be raised: Why is NVT MC not simply alternated with dynamics methods ensuring sampling from the same, that is, canonical distribution? Clearly, Newtonian MD with the widely popular weak-coupling method [109] does not ensure sampling from the proper distribution, but others including LD do. Assuming short momentum autocorrelation times in the presence of significant friction, any error introduced by resetting velocities periodically should be small to negligible. One concern is theoretical in nature and arises from the fact that such an approach cannot be easily cast as a single Markov chain. Another issue might simply be a technical point and relate to considerations outlined in Section 2.2.

We conclude that gradient-based techniques, in particular hybrid MD/MC protocols, are of fundamental importance to the biomacromolecular simulation field due to their universality. They are universal in that they are independent of the details of the system as long as the potential energy function is differentiable. While the implementation challenge of deriving and calculating analytical gradients is nontrivial for certain Hamiltonians, such methods present the most intuitive and straightforward route to introduce correlation into the evolution of the system. We therefore recommend that a gradient-based hybrid method, which rigorously samples the canonical ensemble, be added as an elementary move to available MC software. The aforementioned universality stands in contrast to techniques designed specifically for lattice systems or bead-spring polymer models. Some of those latter techniques and their potential as tools in biomacromolecular MC simulations are discussed next.

2.4.4 Other advanced techniques and their applicability to biomacromolecular systems

The polymer literature yields a variety of specialized move types in particular for lattice homopolymers [110]. Sampling methods like the slithering snake and reptation algorithms (see ref. 111 and references therein) or the original configurational-bias/chain growth algorithms [112,113] were specifically designed for lattice representations. Despite their extension to continuum systems and subsequent improvements [114–122], successful applications to biomacromolecules at an all-atom representation have not been reported. Certainly, the complexity of actual biological heteropolymers eliminates typical assumptions about molecular topology and geometry, which are taken advantage of in these cases.

However, it also needs to be clarified that not all sampling problems can be solved by introducing correlation to the moveset. To illustrate this point, consider Panel B of Figure 4. Instead of an almost barrier-free path connecting the two minima, it is just as well possible for the barrier to be insurmountable. In this case, stepwise MC and MD methodologies are stunted. This is where multicanonical techniques come into play by utilizing data obtained from systems under different conditions, usually different temperatures. A very brief overview of such techniques is given in Section 2.5. Of course, there are multitudinous techniques that drop the requirement to stringently sample from the correct ensemble and can be more easily classified as energy landscape exploration tools [123]. Often used in conjunction with library-based approaches, such methods can potentially be extremely useful in guiding the development of new MC movesets, which specifically capture the intricate correlations needed for efficient sampling using a fundamentally random approach.

2.5 Beyond the canonical ensemble

The common multicanonical techniques such as replica-exchange or simulated tempering have been described and reviewed extensively in different contexts [124]. They interface naturally with MC simulations as they are cast as (biased or unbiased) random walks in terms of a control parameter — usually temperature. They work by exchanging information between the different conditions, thereby allowing increased barrier crossing and quicker convergence of sampling at all conditions of interest.

In addition, a variety of techniques with a narrower focus on enhancing the sampling at a single condition have been developed. There are a variety of techniques employing higher-temperature data to work toward that goal, that is, to reduce the apparent non-ergodicity at the sampling temperature of interest [125–130]. Essentially these techniques can be thought of as forming a continuum as they ultimately rely on similar ideas and/or are based directly on one another. For systems that are of low complexity, the so-called flat-histogram methods [131,132] present another alternative to solve the issue of apparent broken ergodicity. Here, a random walk in energy space is constructed to determine the density of states directly, which then yields thermodynamic quantities. These methods still seem to be restricted to simplified systems as a very recent application to a lattice protein-folding problem demonstrates implicitly [133]. Similarly, the promise of an extension of the method to include a density bias was demonstrated on a discretized protein model [134].

In summary, a wide variety of tools are available to solve canonical sampling problems by using information from different generalized ensembles. Such techniques are truly independent of the underlying MC methodology, whose review in the context of biomacromolecules forms the bulk of this review article. In the next two sections we conclude by presenting a few recent highlights demonstrating the applicability and usefulness of MC sampling to problems of biophysical and/or physicochemical interest.

3. HIGHLIGHTS OF MC SIMULATIONS OF BIOMACROMOLECULES AND OUTLOOK

Undoubtedly, this paragraph needs to be prefaced by the disclaimer that the MC simulation work cited below is only a sample, which is by no means exhaustive. Shaknovich's group has investigated the folding of several biomacromolecules of interest by coupling MC sampling to a simplified Hamiltonian biased toward the native state, that is, a Go model [135–137]. Such studies have been quite feasible due to the better compatibility of MC methods with simple potential energy functions. Another example employing statistical potentials comes from Shental-Bechor et al. [138].

The work of Irbäck deserves special mention where the application of MC methodology to biomacromolecular systems is considered. Irbäck and collaborators [139] recently applied a simple, efficient, knowledge-based implicit solvent model to a variety of biologically relevant problems, ranging from the aggregation of amyloidogenic peptides [140] to the mechanical unfolding of proteins [141]. These are two highlights within a larger body of work [139] that relies exclusively on MC sampling. They demonstrate the potential inherent in combining MC with implicit solvent models.

Similarly, Ulmschneider et al. [26] showed that proper MC sampling can be more efficient than MD for the folding of small peptides. An impressive demonstration of the capability of MC is a recent study on the folding of a transmembrane helix in an implicit membrane environment [142,143]. Vitalis et al. [76] demonstrated that MC simulations can indeed bridge length scales (and hence timescales) inaccessible to conventional dynamics techniques. They simulated the dimer formation of two intrinsically disordered polypeptides and obtained converged associativity data at effective concentrations as low as 100 µM. A dynamics-based approach would have been infeasible in this case due to the limitation of molecules having to diffuse hundreds of angstroms.

De Mori et al. have taken a different approach to take advantage of MC simulations. They used a coarse-grained Hamiltonian to presample phase space in an approximate manner. This is followed by MD simulations starting from representative structures from the most dominantly populated clusters within the MC ensemble. Such a hierarchical strategy was employed to study the folding of a small protein [144] and the oligomer formation of short, amyloidogenic peptides [145].

We wish to reiterate the theme of combining implicit representations of the solvent environment with all-atom models of the biomacromolecules taking advantage of the sampling benefits of MC. Clearly, the boundaries of computer simulation can be pushed to limits that are not easily reached given finite resources. The time is right for the simulation community to participate in the application and development of MC methodology for biomacromolecular systems.

ACKNOWLEDGMENTS

This work was supported by grants MCB 0718924 from the National Science Foundation and RO1 NS056114 from the National Institutes of Health to R.V. Pappu. We are grateful to Xiaoling Wang, Adam Steffen, Nicholas Lyle, Hoang Tran, Tim Williamson, Emma Morrison, and Albert Mao for helpful discussions and data generated using the CAMPARI software package (or earlier versions of this package) that helped guide our thinking regarding Monte Carlo simulations. R.V. Pappu is grateful to Nathan Baker, Alan Grossfield, and Gregory Chirikjian for many helpful discussions over the years.

REFERENCES

1. Plotkin, S.S., Onuchic, J.N. Understanding protein folding with energy landscape theory. Part I: Basic concepts. Q. Rev. Biophys. 2002, 35, 111–67.
2. Plotkin, S.S., Onuchic, J.N. Understanding protein folding with energy landscape theory part II: Quantitative aspects. Q. Rev. Biophys. 2002, 35, 205–86.
3. McCammon, J.A., Gelin, B.R., Karplus, M. Dynamics of folded proteins. Nature 1977, 267, 585–90.
4. Jorgensen, W.L., Tirado-Rives, J. Monte Carlo vs molecular dynamics for conformational sampling. J. Phys. Chem. 1996, 100, 14508–13.
5. Baker, N.A. Improving implicit solvent simulations: a Poisson-centric view. Curr. Opin. Struct. Biol. 2005, 15, 137–43.
6. Chen, J., Brooks Iii, C.L., Khandogin, J. Recent advances in implicit solvent-based methods for biomacromolecular simulations. Curr. Opin. Struct. Biol. 2008, 18, 140–8.
7. Onufriev, A. Implicit solvent models in molecular dynamics simulations: a brief overview. In Annual Reports in Computational Chemistry (eds R.A. Wheeler and D.C. Spellmeyer), Vol. 4, Elsevier, Amsterdam, 2008, pp. 125–37.
8. Frenkel, D., Smit, B. Understanding Molecular Simulation, 2nd ed., Academic Press, San Diego, 2002.
9. Binder, K. Monte Carlo Simulation in Statistical Physics, 3rd ed., Springer-Verlag, Berlin, 1997.
10. Allen, M.P., Tildesley, D.J. Computer Simulation of Liquids, Oxford University Press, Oxford, NY, 1989.
11. Metropolis, N., Rosenbluth, A.W., Rosenbluth, M.N., Teller, A.H., Teller, E. Equation of state calculations by fast computing machines. J. Chem. Phys. 1953, 21, 1087–92.
12. Borovinskiy, A.L., Grosberg, A.Y. Design of toy proteins capable of rearranging conformations in a mechanical fashion. J. Chem. Phys. 2003, 118, 5201–12.
13. Mazenko, G.A. Equilibrium Statistical Mechanics, Wiley Interscience, New York, 2000.
14. Jorgensen, W.L., Tirado-Rives, J. Molecular modeling of organic and biomacromolecular systems using BOSS and MCPRO. J. Comput. Chem. 2005, 26, 1689–700.
15. Hess, B., Kutzner, C., Van Der Spoel, D., Lindahl, E. GROMACS 4: algorithms for highly efficient, load-balanced, and scalable molecular simulation. J. Chem. Theory Comput. 2008, 4, 435–47.
16. Phillips, J.C., Braun, R., Wang, W., Gumbart, J., Tajkhorshid, E., Villa, E., Chipot, C., Skeel, R.D., Kalé, L., Schulten, K. Scalable molecular dynamics with NAMD. J. Comput. Chem. 2005, 26, 1781–802.

17. Ponder, J.W. (2004). TINKER-Software Tools for Molecular Design (http://dasher.wustl.edu/tinker), 4.2 ed.

18. Case, D.A., Darden, T.A., Cheatham, T.E., Simmerling, C.L., Wang, J., Duke, R.E., Luo, R., Crowley, M., Walker, R.C., Zhang, W., Merz, K.M., Wang, B., Hayik, S., Roitberg, A., Seabra, G., Kolossváry, I., Wong, K.F., Paesani, F., Vanicek, J., Wu, X., Brozell, S.R., Steinbrecher, T., Gohlke, H., Yang, L., Tan, C., Mongan, J., Hornak, V., Cui, G., Mathews, D.H., Seetin, M.G., Sagui, C., Babin, V., Kollman, P.A. AMBER 10., University of California, San Francisco, 2008.

19. Brooks, B.R., Bruccoleri, R.E., Olafson, B.D., States, D.J., Swaminathan, S., Karplus, M. CHARMM: a program for macromolecular energy, minimization, and dynamics calculations. J. Comput. Chem. 1983, 4, 187–217.

20. Hu, J., Ma, A., Dinner, A.R. Monte Carlo simulations of biomacromolecules: the MC module in CHARMM. J. Comput. Chem. 2006, 27, 203–16.

21. Martin, M.G. (2008). MCCCS Towhee (http://towhee.sourceforge.net/).

22. Irbäck, A., Mohanty, S. PROFASI: a Monte Carlo simulation package for protein folding and aggregation. J. Comput. Chem. 2006, 27, 1548–55.

23. Kutteh, R., Straatsma, T.P. Molecular dynamics with general holonomic constraints and application to internal coordinate constraints. In Reviews in Computational Chemistry (eds L.D. Lipkowitz and D.B. Boyd), Vol. 12, Wiley, New York, 1998, pp. 75–136.

24. Patriciu, A., Chirikjian, G.S., Pappu, R.V. Analysis of the conformational dependence of mass-metric tensor determinants in serial polymers with constraints. J. Chem. Phys. 2004, 121, 12708–20.

25. Fixman, M. Classical statistical mechanics of constraints: a theorem and application to polymers. Proc. Natl. Acad. Sci. U.S.A. 1974, 71, 3050–53.

26. Ulmschneider, J.P., Ulmschneider, M.B., Di Nola, A. Monte Carlo vs molecular dynamics for all-atom polypeptide folding simulations. J. Phys. Chem. B 2006, 110, 16733–42.

27. Cahill, S., Cahill, M., Cahill, K. On the kinematics of protein folding. J. Comput. Chem. 2003, 24, 1364–70.

28. Karplus, M., Kushick, J.N. Method for estimating the configurational entropy of macromolecules. Macromolecules 1981, 14, 325–32.

29. Fitzgerald, J.E., Jha, A.K., Sosnick, T.R., Freed, K.F. Polypeptide motions are dominated by peptide group oscillations resulting from dihedral angle correlations between nearest neighbors. Biochemistry 2007, 46, 669–82.

30. Vitalis, A., Pappu, R.V. ABSINTH: a new continuum solvation model for simulations of polypeptides in aqueous solutions. J. Comput. Chem. 2009, 30, 673–99.

31. Kaminski, G.A., Friesner, R.A., Tirado-Rives, J., Jorgensen, W.L. Evaluation and reparametrization of the OPLS-AA force field for proteins via comparison with accurate quantum chemical calculations on peptides. J. Phys. Chem. B 2001, 105, 6474–87.

32. Skeel, R.D., Izaguirre, J.A. An impulse integrator for Langevin dynamics. Mol. Phys. 2002, 100, 3885–91.

33. Duan, Y., Wu, C., Chowdhury, S., Lee, M.C., Xiong, G., Zhang, W., Yang, R., Cieplak, P., Luo, R., Lee, T., Caldwell, J., Wang, J., Kollman, P. A point-charge force field for molecular mechanics simulations of proteins based on condensed-phase quantum mechanical calculations. J. Comput. Chem. 2003, 24, 1999–2012.

34. Foloppe, N., MacKerell, A.D. Jr. All-atom empirical force field for nucleic acids: I. Parameter optimization based on small molecule and condensed phase macromolecular target data. J. Comput. Chem. 2000, 21, 86–104.

35. MacKerell, A.D. Jr., Bashford, D., Bellott, M., Dunbrack, R.L. Jr., Evanseck, J.D., Field, M.J., Fischer, S., Gao, J., Guo, H., Ha, S., Joseph-McCarthy, D., Kuchnir, L., Kuczera, K., Lau, F.T.K., Mattos, C., Michnick, S., Ngo, T., Nguyen, D.T., Prodhom, B., Reiher Iii, W.E., Roux, B., Schlenkrich, M., Smith, J.C., Stote, R., Straub, J., Watanabe, M., Wiórkiewicz-Kuczera, J., Yin, D., Karplus, M. All-atom empirical potential for molecular modeling and dynamics studies of proteins. J. Phys. Chem. B 1998, 102, 3586–616.

36. Oostenbrink, C., Villa, A., Mark, A.E., Van Gunsteren, W.F. A biomacromolecular force field based on the free enthalpy of hydration and solvation: the GROMOS force-field parameter sets 53A5 and 53A6. J. Comput. Chem. 2004, 25, 1656–76.

37. Darden, T., York, D., Pedersen, L. Particle mesh Ewald: an $N \cdot \log(N)$ method for Ewald sums in large systems. J. Chem. Phys. 1993, 98, 10089–92.
38. Mallik, B., Masunov, A., Lazaridis, T. Distance and exposure dependent effective dielectric function. J. Comput. Chem. 2002, 23, 1090–9.
39. Michel, J., Taylor, R.D., Essex, J.W. Efficient generalized born models for Monte Carlo simulations. J. Chem. Theory Comput. 2006, 2, 732–9.
40. Nilmeier, J., Jacobson, M. Multiscale Monte Carlo sampling of protein sidechains: application to binding pocket flexibility. J. Chem. Theory Comput. 2008, 4, 835–46.
41. Gelb, L.D. Monte Carlo simulations using sampling from an approximate potential. J. Chem. Phys. 2003, 118, 7747–50.
42. Hetényi, B., Bernacki, K., Berne, B.J. Multiple "time step" Monte Carlo. J. Chem. Phys. 2002, 117, 8203–7.
43. Essmann, U., Perera, L., Berkowitz, M.L., Darden, T., Lee, H., Pedersen, L.G. A smooth particle mesh Ewald method. J. Chem. Phys. 1995, 103, 8577–93.
44. Bernacki, K., Hetényi, B., Berne, B.J. Multiple "time step" Monte Carlo simulations: application to charged systems with Ewald summation. J. Chem. Phys. 2004, 121, 44–50.
45. Ewald, P.P. Die Berechnung optischer und elektrostatischer Gitterpotentiale. Annalen der Physik 1921, 369, 253–87.
46. Brush, S.G., Sahlin, H.L., Teller, E. Monte-Carlo study of a one-component plasma. I. J. Chem. Phys. 1966, 45, 2102–18.
47. Hansen, J.P. Statistical mechanics of dense ionized matter. I. Equilibrium properties of the classical one-component plasma. Phys. Rev. A 1973, 8, 3096–109.
48. Adams, D.J., Dubey, G.S. Taming the Ewald sum in the computer simulation of charged systems. J. Comput. Phys. 1987, 72, 156–76.
49. Toukmaji, A.Y., Board, J.A. Jr. Ewald summation techniques in perspective: a survey. Comput. Phys. Commun. 1996, 95, 73–92.
50. Tironi, I.G., Sperb, R., Smith, P.E., Van Gunsteren, W.F. A generalized reaction field method for molecular dynamics simulations. J. Chem. Phys. 1995, 102, 5451–9.
51. Hockney, R.W., Eastwood, J.W. Computer Simulations Using Particles, McGraw-Hill, New York, 1981.
52. Verlet, L. Computer "experiments" on classical fluids. I. Thermodynamical properties of Lennard-Jones molecules. Phys. Rev. 1967, 159, 98–103.
53. Murthy, V.L., Srinivasan, R., Draper, D.E., Rose, G.D. A complete conformational map for RNA. J. Mol. Biol. 1999, 291, 313–27.
54. Lal, M. 'Monte Carlo' computer simulation of chain molecules. I. Mol. Phys. 1969, 17, 57–64.
55. Dunbrack, R.L. Jr. Rotamer libraries in the 21st century. Curr. Opin. Struct. Biol. 2002, 12, 431–40.
56. Ho, B.K., Coutsias, E.A., Seok, C., Dill, K.A. The flexibility in the proline ring couples to the protein backbone. Protein Sci. 2005, 14, 1011–8.
57. Vitagliano, L., Berisio, R., Mastrangelo, A., Mazzarella, L., Zagari, A. Preferred proline puckerings in cis and trans peptide groups: implications for collagen stability. Protein Sci. 2001, 10, 2627–32.
58. Saenger, W. Principles of Nucleic Acid Structure, 1st ed., Springer, New York, 1984.
59. Sklenar, H., Wüstner, D., Rohs, R. Using internal and collective variables in Monte Carlo simulations of nucleic acid structures: chain breakage/closure algorithm and associated Jacobians. J. Comput. Chem. 2006, 27, 309–15.
60. Binder, K., Baschnagel, J., Paul, W. Glass transition of polymer melts: test of theoretical concepts by computer simulation. Prog. Polymer Sci. (Oxford) 2003, 28, 115–72.
61. Tanaka, M., Grosberg, A.Y., Tanaka, T. Molecular dynamics simulations of polyampholytes. Langmuir 1999, 15, 4052–5.
62. Takada, S., Wolynes, P.G. Glassy dynamics of random heteropolymers. Prog. Theor. Phys. Suppl. 1997, 49–52.
63. Vitalis, A., Wang, X., Pappu, R.V. Quantitative characterization of intrinsic disorder in polyglutamine: insights from analysis based on polymer theories. Biophys. J. 2007, 93, 1923–37.

64. Frisch, T., Verga, A. Slow relaxation and solvent effects in the collapse of a polymer. Phys. Rev. E Stat. Nonlin. Soft Matter Phys. 2002, 66, 041807.
65. Ziv, G., Thirumalai, D., Haran, G. Collapse transition in proteins. Phys. Chem. Chem. Phys. 2009, 11, 83–93.
66. Straub, J.E., Thirumalai, D. Exploring the energy landscape in proteins. Proc. Natl. Acad. Sci. U.S.A. 1993, 90, 809–13.
67. Van Gunsteren, W.F., Bakowies, D., Baron, R., Chandrasekhar, I., Christen, M., Daura, X., Gee, P., Geerke, D.P., Glättli, A., Hünenberger, P.H., Kastenholz, M.A., Oostenbrink, C., Schenk, M., Trzesniak, D., Van Der Vegt, N.F.A., Yu, H.B. Biomacromolecular modeling: goals, problems, perspectives. Angew. Chem. Int. Ed. 2006, 45, 4064–92.
68. Hess, B. Convergence of sampling in protein simulations. Phys. Rev. E Stat. Nonlin. Soft Matter Phys. 2002, 65, 031910.
69. Frickenhaus, S., Kannan, S., Zacharias, M. Efficient evaluation of sampling quality of molecular dynamics simulations by clustering of dihedral torsion angles and sammon mapping. J. Comput. Chem. 2009, 30, 479–92.
70. Son, W.J., Jang, S., Shin, S. A simple method of estimating sampling consistency based on free energy map distance. J. Mol. Graph. Model. 2008, 27, 321–5.
71. Monticelli, L., Sorin, E.J., Tieleman, D.P., Pande, V.S., Colombo, G. Molecular simulation of multistate peptide dynamics: a comparison between microsecond timescale sampling and multiple shorter trajectories. J. Comput. Chem. 2008, 29, 1740–52.
72. Karney, C.F.F. Quaternions in molecular modeling. J. Mol. Graph. Model. 2007, 25, 595–604.
73. Franklin, J., Doniach, S. Adaptive time stepping in biomacromolecular dynamics. J. Chem. Phys. 2005, 123, 124909.
74. Barth, E., Leimkuhler, B., Reich, S. Time-reversible variable-stepsize integrator for constrained dynamics. SIAM J. Sci. Comput. 1999, 21, 1027–44.
75. Izaguirre, J.A., Catarello, D.P., Wozniak, J.M., Skeel, R.D. Langevin stabilization of molecular dynamics. J. Chem. Phys. 2001, 114, 2090–8.
76. Vitalis, A., Wang, X., Pappu, R.V. Atomistic simulations of the effects of polyglutamine chain length and solvent quality on conformational equilibria and spontaneous homodimerization. J. Mol. Biol. 2008, 384, 279–97.
77. Sokal, A.D. Monte Carlo methods for the self-avoiding walk. In Monte Carlo and Molecular Dynamics Simulations in Polymer Science (ed. K. Binder), Oxford University Press, New York, 1995, pp. 47–124.
78. Podtelezhnikov, A.A., Wild, D.L. Exhaustive Metropolis Monte Carlo sampling and analysis of polyalanine conformations adopted under the influence of hydrogen bonds. Protein. Struct. Funct. Genet. 2005, 61, 94–104.
79. Go, N., Scheraga, H.A. Ring closure and local conformational deformations of chain molecules. Macromolecules 1970, 3, 178–87.
80. Bruccoleri, R.E., Karplus, M. Chain closure with bond angle variations. Macromolecules 1985, 18, 2767–73.
81. Ulmschneider, J.P., Jorgensen, W.L. Monte Carlo backbone sampling for polypeptides with variable bond angles and dihedral angles using concerted rotations and a Gaussian bias. J. Chem. Phys. 2003, 118, 4261–71.
82. Dodd, L.R., Boone, T.D., Theodorou, D.N. A concerted rotation algorithm for atomistic Monte Carlo simulation of polymer melts and glasses. Mol. Phys. 1993, 78, 961–96.
83. Hoffmann, D., Knapp, E.W. Polypeptide folding with off-lattice Monte Carlo dynamics: the method. Eur. Biophys. J. 1996, 24, 387–403.
84. Dinner, A.R. Local deformations of polymers with nonplanar rigid main-chain internal coordinates. J. Comput. Chem. 2000, 21, 1132–44.
85. Mezei, M. Efficient Monte Carlo sampling for long molecular chains using local moves, tested on a solvated lipid bilayer. J. Chem. Phys. 2003, 118, 3874–9.
86. Favrin, G., Irbäck, A., Sjunnesson, F. Monte Carlo update for chain molecules: biased Gaussian steps in torsional space. J. Chem. Phys. 2001, 114, 8154–8.

87. Coutsias, E.A., Seok, C., Jacobson, M.P., Dill, K.A. A kinematic view of loop closure. J. Comput. Chem. 2004, 25, 510–28.
88. Kolodny, R., Guibas, L., Levitt, M., Koehl, P. Inverse kinematics in biology: the protein loop closure problem. Int. J. Robot. Res. 2005, 24, 151–63.
89. Swendsen, R.H., Wang, J.S. Nonuniversal critical dynamics in Monte Carlo simulations. Phys. Rev. Lett. 1987, 58, 86–8.
90. Wolff, U. Collective Monte Carlo updating for spin systems. Phys. Rev. Lett. 1989, 62, 361–4.
91. Luijten, E. Introduction to cluster Monte Carlo algorithms. In Computer Simulations in Condensed Matter Systems: From Materials to Chemical Biology (eds M. Ferrario, G. Ciccotti and K. Binder), Vol. 1–703, Springer, Berlin, 2006, pp. 13–38.
92. Liu, J., Luijten, E. Rejection-free geometric cluster algorithm for complex fluids. Phys. Rev. Lett. 2004, 92, 035504.
93. Dress, C., Krauth, W. Cluster algorithm for hard spheres and related systems. J. Phys. A Gen. Phys. 1995, 28, L597–601.
94. Whitelam, S., Geissler, P.L. Avoiding unphysical kinetic traps in Monte Carlo simulations of strongly attractive particles. J. Chem. Phys. 2007, 127, 154101.
95. Bhattacharyay, A., Troisi, A. Self-assembly of sparsely distributed molecules: an efficient cluster algorithm. Chem. Phys. Lett. 2008, 458, 210–3.
96. Liu, J., Luijten, E. Generalized geometric cluster algorithm for fluid simulation. Phys. Rev. E Stat. Nonlin. Soft Matter Phys. 2005, 71, 066701.
97. Maggs, A.C. Multiscale Monte Carlo algorithm for simple fluids. Phys. Rev. Lett. 2006, 97, 197802.
98. Pangali, C., Rao, M., Berne, B.J. On a novel Monte Carlo scheme for simulating water and aqueous solutions. Chem. Phys. Lett. 1978, 55, 413–7.
99. Rossky, P.J., Doll, J.D., Friedman, H.L. Brownian dynamics as smart Monte Carlo simulation. J. Chem. Phys. 1978, 69, 4628–33.
100. Heyes, D.M., Branka, A.C. Monte Carlo as Brownian dynamics. Mol. Phys. 1998, 94, 447–54.
101. Kotelyanskii, M.J., Suter, U.W. A dynamic Monte Carlo method suitable for molecular simulations. J. Chem. Phys. 1992, 96, 5383–8.
102. Duane, S., Kennedy, A.D., Pendleton, B.J., Roweth, D. Hybrid Monte Carlo. Phys. Lett. B 1987, 195, 216–22.
103. Kennedy, A.D., Pendleton, B. Acceptances and autocorrelations in hybrid Monte Carlo. Nucl. Phys. B (Proc. Suppl.) 1991, 20, 118–21.
104. Mackenzie, P.B. An improved hybrid Monte Carlo method. Phys. Lett. B 1989, 226, 369–71.
105. Faller, R., De Pablo, J.J. Constant pressure hybrid molecular dynamics — Monte Carlo simulations. J. Chem. Phys. 2002, 116, 55–9.
106. Izaguirre, J.A., Hampton, S.S. Shadow hybrid Monte Carlo: an efficient propagator in phase space of macromolecules. J. Comput. Phys. 2004, 200, 581–604.
107. Akhmatskaya, E., Bou-Rabee, N., Reich, S. A comparison of generalized hybrid Monte Carlo methods with and without momentum flip. J. Comput. Phys. 2009, 228, 2256–65.
108. Kennedy, A.D. Higher order hybrid Monte Carlo. Nucl. Phys. B (Proc. Suppl.) 1989, 9, 457–62.
109. Berendsen, H.J.C., Postma, J.P.M., Van Gunsteren, W.F., Di Nola, A., Haak, J.R. Molecular dynamics with coupling to an external bath. J. Chem. Phys. 1984, 81, 3684–90.
110. Kremer, K., Binder, K. Monte Carlo simulation of lattice models for macromolecules. Comput. Phys. Rep. 1988, 7, 259–310.
111. Nidras, P.P., Brak, R. New Monte Carlo algorithms for interacting self-avoiding walks. J. Phys. A Math. Gen. 1997, 30, 1457–69.
112. Siepmann, J.I., Frenkel, D. Configurational bias Monte Carlo: a new sampling scheme for flexible chains. Mol. Phys. 1992, 75, 59–70.
113. O'Toole, E.M., Panagiotopoulos, A.Z. Monte Carlo simulation of folding transitions of simple model proteins using a chain growth algorithm. J. Chem. Phys. 1992, 97, 8644–52.
114. Frenkel, D., Mooij, G.C.A.M., Smit, B. Novel scheme to study structural and thermal properties of continuously deformable molecules. J. Phys. Condens. Matter 1992, 4, 3053–76.
115. Vendruscolo, M. Modified configurational bias Monte Carlo method for simulation of polymer systems. J. Chem. Phys. 1997, 106, 2970–6.

116. Vlugt, T.J.H., Martin, M.G., Smit, B., Siepmann, J.I., Krishna, R. Improving the efficiency of the configurational-bias Monte Carlo algorithm. Mol. Phys. 1998, 94, 727–33.
117. Zhang, J., Kou, S.C., Liu, J.S. Biopolymer structure simulation and optimization via fragment regrowth Monte Carlo. J. Chem. Phys. 2007, 126, 225101.
118. Escobedo, F.A., De Pablo, J.J. Extended continuum configurational bias Monte Carlo methods for simulation of flexible molecules. J. Chem. Phys. 1995, 102, 2636–52.
119. Houdayer, J. The wormhole move: a new algorithm for polymer simulations. J. Chem. Phys. 2002, 116, 1783–7.
120. Laso, M., Karayiannis, N.C., Müller, M. Min-map bias Monte Carlo for chain molecules: biased Monte Carlo sampling based on bijective minimum-to-minimum mapping. J. Chem. Phys. 2006, 125, 164108.
121. Grassberger, P. Pruned-enriched Rosenbluth method: simulations of θ polymers of chain length up to 1 000 000. Phys. Rev. E –Stat. Phys. Plasmas Fluids Relat. Interdiscip. Topics 1997, 56, 3682–93.
122. Consta, S., Vlugt, T.J.H., Hoeth, J.W., Smit, B., Frenkel, D. Recoil growth algorithm for chain molecules with continuous interactions. Mol. Phys. 1999, 97, 1243–54.
123. Prentiss, M.C., Wales, D.J., Wolynes, P.G. Protein structure prediction using basin-hopping. J. Chem. Phys. 2008, 128, 225106.
124. Sugita, Y., Mitsutake, A., Okamoto, Y. Generalized-ensemble algorithms for protein folding simulations. In Rugged Free Energy Landscapes (ed. W. Janke), Vol. 736, Springer, Berlin, 2008, pp. 369–407.
125. Frantz, D.D., Freeman, D.L., Doll, J.D. Reducing quasi-ergodic behavior in Monte Carlo simulations by J-walking: applications to atomic clusters. J. Chem. Phys. 1990, 93, 2769–84.
126. Zhou, R., Berne, B.J. Smart walking: a new method for Boltzmann sampling of protein conformations. J. Chem. Phys. 1997, 107, 9185–96.
127. Xu, H., Berne, B.J. Multicanonical jump walking: a method for efficiently sampling rough energy landscapes. J. Chem. Phys. 1999, 110, 10299–306.
128. Brown, S., Head-Gordon, T. Cool walking: a new Markov chain Monte Carlo sampling method. J. Comput. Chem. 2003, 24, 68–76.
129. Andricioaei, I., Straub, J.E., Voter, A.F. Smart darting Monte Carlo. J. Chem. Phys. 2001, 114, 6994–7000.
130. Nigra, P., Freeman, D.L., Doll, J.D. Combining smart darting with parallel tempering using Eckart space: application to Lennard-Jones clusters. J. Chem. Phys. 2005, 122, 1.
131. Wang, F., Landau, D.P. Efficient, multiple-range random walk algorithm to calculate the density of states. Phys. Rev. Lett. 2001, 86, 2050.
132. Shell, M.S., Debenedetti, P.G., Panagiotopoulos, A.Z. Generalization of the Wang-Landau method for off-lattice simulations. Phys. Rev. E Stat. Nonlin. Soft Matter Phys. 2002, 66, 056703.
133. Wüst, T., Landau, D.P. The HP model of protein folding: a challenging testing ground for Wang-Landau sampling. Comput. Phys. Commun. 2008, 179, 124–7.
134. Thomas, G.L., Sessions, R.B., Parker, M.J. Density guided importance sampling: application to a reduced model of protein folding. Bioinformatics 2005, 21, 2839–43.
135. Shimada, J., Kussell, E.L., Shakhnovich, E.I. The folding thermodynamics and kinetics of crambin using an all-atom Monte Carlo simulation. J. Mol. Biol. 2001, 308, 79–95.
136. Shimada, J., Shakhnovich, E.I. The ensemble folding kinetics of protein G from an all-atom Monte Carlo simulation. Proc. Natl. Acad. Sci. U.S.A. 2002, 99, 11175–80.
137. Nivón, L.G., Shakhnovich, E.I. All-atom Monte Carlo simulation of GCAA RNA folding. J. Mol. Biol. 2004, 344, 29–45.
138. Shental-Bechor, D., Kirca, S., Ben-Tal, N., Haliloglu, T. Monte Carlo studies of folding, dynamics, and stability in α-helices. Biophys. J. 2005, 88, 2391–402.
139. Irbäck, A. Protein folding, unfolding and aggregation studied using an all-atom model with a simplified interaction potential. In Rugged Free Energy Landscapes (ed. W. Janke), Vol. 736, Springer, Berlin, 2008, pp. 269–91.
140. Li, D.W., Mohanty, S., Irbäck, A., Huo, S. Formation and growth of oligomers: a Monte Carlo study of an amyloid tau fragment. PLoS Comput. Biol. 2008, 4, e1000238.

141. Mitternacht, S., Luccioli, S., Torcini, A., Imparato, A., Irbäck, A. Changing the mechanical unfolding pathway of FnIII10 by tuning the pulling strength. Biophys. J. 2009, 96, 429–41.
142. Ulmschneider, J.P., Ulmschneider, M.B., Di Nola, A. Monte Carlo folding of trans-membrane helical peptides in an implicit generalized Born membrane. Protein. Struct. Funct. Genet. 2007, 69, 297–308.
143. Ulmschneider, J.P., Ulmschneider, M.B. Folding simulations of the transmembrane helix of virus protein U in an implicit membrane model. J. Chem. Theory Comput. 2007, 3, 2335–46.
144. De Mori, G.M.S., Colombo, G., Micheletti, C. Study of the villin headpiece folding dynamics by combining coarse-grained Monte Carlo evolution and all-atom molecular dynamics. Protein. Struct. Funct. Genet. 2005, 58, 459–71.
145. Meli, M., Morra, G., Colombo, G. Investigating the mechanism of peptide aggregation: insights from mixed Monte Carlo-molecular dynamics simulations. Biophys. J. 2008, 94, 4414–26.

Section 2
Simulation Methodologies

Section Editor: Carlos Simmerling

Department of Chemistry
State University of New York
Stony Brook, NY 11794-3400
USA

CHAPTER 4

Accelerated Molecular Dynamics Methods: Introduction and Recent Developments

Danny Perez[1], **Blas P. Uberuaga**[2], **Yunsic Shim**[3], **Jacques G. Amar**[3] and **Arthur F. Voter**[1]

Abstract

Because of its unrivaled predictive power, the molecular dynamics (MD) method is widely used in theoretical chemistry, physics, biology, materials science, and engineering. However, due to computational cost, MD simulations can only be used to directly simulate dynamical processes over limited timescales (e.g., nanoseconds or at most a few microseconds), even though the simulation of nonequilibrium processes can often require significantly longer timescales, especially when they involve thermal activation. In this paper, we present an introduction to accelerated molecular dynamics, a class of methods aimed at extending the timescale range of molecular dynamics, sometimes up to seconds or more. The theoretical foundations underpinning the different methods (parallel replica

[1] Theoretical Division, Los Alamos National Laboratory, Los Alamos, NM, USA
[2] Materials Science and Technology Division, Los Alamos National Laboratory, Los Alamos, NM, USA
[3] Department of Physics and Astronomy, University of Toledo, Toledo, OH, USA

Annual Reports in Computational Chemistry, Volume 5
ISSN: 1574-1400, DOI 10.1016/S1574-1400(09)00504-0

dynamics, hyperdynamics, and temperature-accelerated dynamics) are first discussed. We then discuss some applications and recent advances, including super-state parallel replica dynamics, self-learning hyperdynamics, and spatially parallel temperature-accelerated dynamics.

Keywords: infrequent events; transition-state theory; accelerated dynamics; hyperdynamics; parallel-replica dynamics; temperature-accelerated dynamics; molecular dynamics; bond-boost hyperdynamics; parallel-accelerated dynamics; Cu(100)

A long-standing limitation in the use of molecular dynamics (MD) simulation is that it can only be applied directly to processes that take place on very short timescales: typically nanoseconds if empirical potentials are employed, or picoseconds if we rely on electronic structure methods. Many processes of interest in chemistry, biochemistry, and materials science require study over microseconds and beyond, due to either the natural timescale for the evolution or the duration of the experiment of interest. The dynamics on these timescales is typically characterized by infrequent-event transitions, from state to state, usually involving an energy barrier. There is a long and venerable tradition in chemistry of using transition-state theory (TST) [1–3] to compute rate constants directly for these kinds of activated processes. If needed, dynamical corrections to the TST rate, and even quantum corrections, can be computed to achieve an accuracy suitable for the problem at hand. These rate constants then allow us to understand the system behavior on longer timescales than we can directly reach with MD. For complex systems with many reaction paths, the TST rates can be fed into a stochastic simulation procedure such as kinetic Monte Carlo (KMC) [4], and a direct simulation of the advance of the system through its possible states can be obtained in a probabilistically exact way.

A problem that has become more evident in recent years, however, is that for many systems of interest there is a complexity that makes it difficult, if not impossible, to determine all the relevant reaction paths to which TST should be applied. This is a serious issue, as omitted transition pathways can have uncontrollable consequences on the simulated long-time kinetics.

Over the past decade or so, we have been developing a new class of methods for treating the long-time dynamics in these complex, infrequent-event systems. Rather than trying to guess in advance what reaction pathways may be important, we return instead to an MD treatment, in which the trajectory itself finds an appropriate way to escape from each state of the system. Since a direct integration of the trajectory would be limited to nanoseconds, while we are seeking to follow the system for much longer times, we modify the dynamics so that the first escape will happen much more quickly, thereby accelerating the dynamics. The key is to design the modified dynamics in a way that does as little damage as possible to the probability for escaping along a given pathway — i.e., we try to preserve the relative rate constants for the different possible escape paths out of the state. We can then use this modified dynamics to follow the system from state to state, reaching much longer times than we could reach with direct MD. The dynamics within any one state may no longer be meaningful, but the state-to-state dynamics, in the best case (as we discuss below), can be exact. We have developed three methods in this *accelerated molecular dynamics* (AMD) class, in each case appealing to TST, either implicitly or explicitly, to design the modified dynamics. Each of the methods has its own advantages, and we and others have applied these methods to a wide range of problems. The purpose of this article is to give the reader a brief introduction to how these methods work, and discuss some of the recent developments that have been made to improve their power and applicability. Note that this brief review does not claim to be exhaustive: various other methods aiming at similar goals have been proposed in the literature. For the sake of brevity, our focus will exclusively be on the methods developed by our group.

This paper is organized as follows: Sections 1, 2, and 3 discuss the basic theoretical foundations underlying the three main AMD methods, namely parallel-replica dynamics, hyperdynamics, and temperature-accelerated dynamics (TAD), respectively. In Section 4, we present some recent improvements of the original AMD methods, allowing them to tackle more complex, multi-timescale, systems (Section 4.1), to tune themselves automatically to the system at hand (Section 4.2) and to efficiently simulate systems of large sizes (Section 4.3). We then conclude by discussing current and forthcoming challenges that need to be addressed to further extend the applicability and performance of the AMD methods.

1. PARALLEL-REPLICA DYNAMICS

The parallel-replica method [5] is perhaps the least glamorous of the AMD methods, but is, in many cases, the most powerful. It is also the most accurate AMD method, assuming only first-order kinetics (exponential decay); i.e., for any trajectory that has been in a state long enough to have lost its memory of how it entered the state (longer than the correlation time τ_{corr}, the time after which the system is effectively sampling a stationary distribution restricted to the current state), the probability distribution function for the time of the next escape from that state is given by

$$p(t) = ke^{-kt} \tag{1}$$

where k is the rate constant for escape from the state. Parallel-replica allows for the temporal parallelization of the state-to-state dynamics of such a system on M processors. This is to be contrasted with standard parallelizations of MD simulations in which spatial decomposition schemes are used.

We sketch the derivation here. For a state with total escape rate k which is simultaneously explored on M processors, the effective escape rate for the first escape of *any* replica is Mk. If the simulation time accumulated on one processor is t, the total time on the M processors will then be $t_{sum} = Mt$. Thus, using a simple change of variable, $p(t)$ can be written as:

$$p(t)dt = Mke^{-Mkt}dt \tag{2}$$

$$= ke^{-kt_{sum}}dt_{sum} \tag{3}$$

$$= p(t_{sum})dt_{sum} \tag{4}$$

implying that the probability to leave the state per unit MD time is the same whether the simulation is run on one or M processors. While this derivation applies for processors of equal speed, the same conclusion can be shown to be valid if a heterogeneous set of processors is instead used; see ref. 5.

Figure 1 shows a schematic of the algorithm. Starting with a system in a particular state, it is replicated on each of the M processors. Each replica is evolved forward with independent thermostats for a time $\Delta t_{deph} \geq \tau_{corr}$ to eliminate correlations between replicas, a stage referred to as dephasing. After dephasing, each processor carries out an independent constant-temperature MD trajectory, together exploring phase space within the particular basin M times faster than a single trajectory would. Once a transition is detected on any processor, all processors are stopped. The simulation clock is then advanced by t_{sum}, the accumulated trajectory time summed over all replicas until the transition occurred.

The parallel-replica method also correctly accounts for correlated dynamical events (there is no requirement that the system obeys TST), unlike the other AMD methods. This is accomplished by allowing the trajectory that made the transition to continue for a further amount of time $\Delta t_{corr} \geq \tau_{corr}$, during which recrossings or follow-on events may occur. The simulation clock is then advanced by Δt_{corr}, the new state is replicated on all processors, and the whole process is repeated.

The computational efficiency of the method is limited by both the dephasing stage, which does not advance the system clock, and the correlated-event stage,

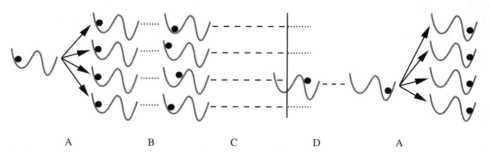

        A   B   C   D   A

Figure 1   Schematic illustration of the parallel-replica method. The four steps, described in the text, are (A) replication of the system into M copies, (B) dephasing of the replicas, (C) propagation of independent trajectories until a transition is detected in any of the replicas, and (D) brief continuation of the transitioning trajectory to allow for correlated events such as recrossings or follow-on transitions to other states. The resulting configuration is then replicated, beginning the process again. Reprinted, with permission, from ref. 6. Copyright 2002 by Annual Reviews. www.annualreviews.org

during which only one processor accumulates time. (This is illustrated schematically in Figure 1, where dashed line trajectories advance the simulation clock but dotted line trajectories do not.) Thus, the overall efficiency will be high when

$$\frac{1}{kM} \gg \Delta t_{\text{deph}} + \Delta t_{\text{corr}} \tag{5}$$

An extension to parallel-replica allows the method to be applied to driven systems. To result in valid dynamics, the drive rate must be slow enough that at any given time the rates for the different pathways in the system depend only on the instantaneous configuration of the system [7].

Parallel-replica dynamics has been successfully applied to a number of different problems, including the diffusion of H_2 in crystalline C_{60} [8], the pyrolysis of hexadecane [9], the diffusion of defects in plutonium [10], the transformation of voids into stacking fault tetrahedra in FCC metals [11], the stretching of carbon nanotubes [7], grain boundary sliding in Cu [12], the diffusion of Li through a polymer matrix [13], the fracture process of metals [14], and the folding dynamics of small proteins [15]. As parallel-computing environments become more common, the parallel-replica method will become an increasingly important tool for the exploration of complex systems.

2. HYPERDYNAMICS

Another possible avenue to accelerate the state-to-state evolution of a system of interest is to construct an auxiliary system in such a way that the dynamics of the latter are faster than that of the former while enforcing that one maps onto the other by a suitable renormalization of time. Hyperdynamics [16] realizes this objective by building on the concept of importance sampling [17,18] and extending it into the time domain. In this approach, the auxiliary system is obtained by adding a nonnegative *bias* potential $\Delta V_b(\mathbf{r})$ to the potential of the original system $V(\mathbf{r})$ so that the height of the barriers between different states is reduced, as schematically shown in Figure 2. The relationship between the dynamical evolution of the original and biased systems is recovered using TST. Indeed, according to TST, the rate of escape of the original system out of a given state A is given by

$$k_{A\rightarrow}^{\text{TST}} = \langle |v_A| \delta_A(\mathbf{r}) \rangle_A \tag{6}$$

where $\delta_A(\mathbf{r})$ is a Dirac delta function centered on the separatrix hypersurface between state A (i.e., the hypersurface is at $\mathbf{r} = 0$) and the neighboring states, v_A the velocity normal to it, and $\langle P \rangle_A$ the canonical ensemble average of a quantity P for a system confined to state A. By standard importance sampling manipulations, the last equation can be recast in a form where the averages are obtained on the *biased* potential instead. We find:

$$k_{A\rightarrow}^{\text{TST}} = \frac{\langle |v_A| \delta_A(\mathbf{r}) e^{\beta \Delta V_b(\mathbf{r})} \rangle_{A_b}}{\langle e^{\beta \Delta V_b(\mathbf{r})} \rangle_{A_b}} \tag{7}$$

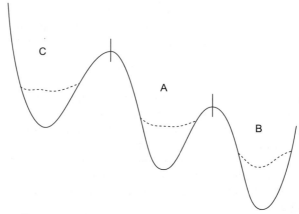

Figure 2 Schematic illustration of the hyperdynamics method. A bias potential ($\Delta V(r)$) is added to the original potential ($V(\mathbf{r})$, solid line). Provided that $\Delta V(r)$ meets certain conditions, primarily that it be zero at the dividing surfaces between states, a trajectory on the biased potential surface ($V(\mathbf{r}) + \Delta V(\mathbf{r})$, dashed line) escapes more rapidly from each state without corrupting the relative escape probabilities. The accelerated time is estimated as the simulation proceeds.

where $\beta = 1/k_B T$ and k_B is the Boltzmann constant. If we impose the condition that the bias potential vanish at the separatrix, the last equation can be rewritten as

$$k_{A\to}^{\mathrm{TST}} = \frac{\langle |v_A|\delta(\mathbf{r})\rangle_{A_b}}{\langle e^{\beta \Delta V_b(\mathbf{r})}\rangle_{A_b}} \tag{8}$$

This result is very appealing since the relative rates of escapes from A to other states are invariant under the addition of the bias potential, i.e.,

$$\frac{k_{A_b\to B}^{\mathrm{TST}}}{k_{A_b\to C}^{\mathrm{TST}}} = \frac{k_{A\to B}^{\mathrm{TST}}}{k_{A\to C}^{\mathrm{TST}}} \tag{9}$$

Thus, the state-to-state dynamics on the biased potential is equivalent to that on the original potential as long as the time is renormalized to account for the uniform relative increase of all the rates introduced by the biased potential. This renormalization is in practice obtained by multiplying the MD timestep Δt_{MD} by the inverse Boltzmann factor for the bias potential, so that n MD timesteps on the biased potential are equivalent to an elapsed time of

$$t_{\mathrm{hyper}} = \sum_{j=1}^{n} \Delta t_{\mathrm{MD}} e^{\Delta V(r(t_j))/k_B T} \tag{10}$$

on the original potential. This renormalization can be shown to be exact in the long-time limit. The overall computational speedup for hyperdynamics is simply given by the average boost factor

$$\text{boost(hyperdynamics)} = \frac{t_{\text{hyper}}}{t_{\text{MD}}} \tag{11}$$

divided by the relative extra cost of calculating the bias potential and associated forces.

Thus, if both the original and biased systems obey TST so that the above-mentioned derivation holds, hyperdynamics can provide considerable acceleration compared to direct-MD simulations. However, in practice, the applicability of hyperdynamics is limited by the availability of low-overhead bias potentials. Indeed, while some different forms have been proposed in the last few years, often they are computationally expensive, tailored to a limited class of systems or built on sets of restrictive assumptions about the nature of the separatrix. The main challenge, which is the subject of active research in different groups, thus remains the construction of bias potentials that are simple, efficient, generic, and transferable. We present below one recent advance in this area.

Despite the aforementioned difficulties, hyperdynamics has been successfully applied to a variety of systems, including desorption of organic molecules from graphitic substrates [19], surface diffusion of metallic clusters [20], heteroepitaxial growth [21], and the dynamics of biomolecules [22].

3. TEMPERATURE-ACCELERATED DYNAMICS

One natural way of speeding up the dynamics of a system is to simply raise the temperature. However, while the rates of processes will increase with higher temperatures, the relative probabilities of different events occurring will be different than at the original temperature of interest. Correcting for this reordering is the basic idea behind TAD [23]. In TAD, transitions are sped up by increasing the temperature to some T_{high}, but transitions that should not have occurred at the original temperature T_{low} are filtered out. The TAD method assumes that the system obeys harmonic TST and, as a result, is more approximate than the other AMD methods. However, for many applications, especially in solids, this additional approximation is acceptable.

In each basin, the system is evolved at T_{high}. When a transition is detected, the saddle point for that transition is found. The trajectory is then reflected back into the basin and continued. This procedure is referred to as "basin constrained molecular dynamics" (BCMD). During the BCMD, a list of escape paths and escape times at T_{high} for each pathway is generated. Assuming that harmonic TST holds, and knowing the saddle point energy for the transition, we can then extrapolate each escape time observed at T_{high} to obtain a corresponding escape time at T_{low}. This extrapolation, which does not require knowing the preexponential factor, can be illustrated graphically in an Arrhenius-style plot ($\ln(1/t)$ vs. $1/T$), as shown in Figure 3. The time for each event seen at T_{high}

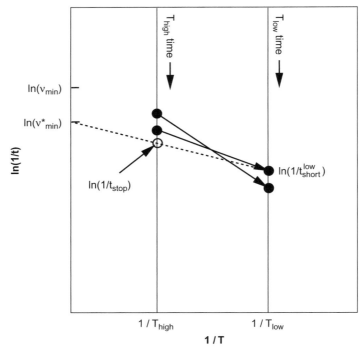

Figure 3 Schematic illustration of the temperature-accelerated dynamics method. Progress of the high-temperature trajectory can be thought of as moving down the vertical timeline at $1/T_{high}$. For each transition detected during the run, the trajectory is reflected back into the basin, the saddle point is found, and the time of the transition (solid dot on left timeline) is transformed (arrow) into a time on the low-temperature timeline. Plotted in this Arrhenius-like form, this transformation is a simple extrapolation along a line whose slope is the negative of the barrier height for the event. The dashed termination line connects the shortest-time transition recorded so far on the low-temperature timeline with the confidence-modified minimum preexponential ($v_{min}^{*} = v_{min}/\ln(1/\delta)$) on the y-axis. The intersection of this line with the high-T timeline gives the time (t_{stop}, open circle) at which the trajectory can be terminated. With confidence $1-\delta$, we can say that any transition observed after t_{stop} could only extrapolate to a shorter time on the low-T timeline if it had a preexponential factor lower than v_{min}.

extrapolated to T_{low} is

$$t_{low} = t_{high}e^{E_a(\beta_{low}-\beta_{high})} \tag{12}$$

where $\beta = 1/k_B T$ and E_a is the activation energy.

As the BCMD is continued, a new shorter-time event may be discovered. With the additional assumption that there is a minimum preexponential factor, v_{min}, which bounds from below all the preexponential factors in the system, we can define a time at which the BCMD trajectory can be stopped. This time has the property that the probability any transition observed later would replace the first transition at T_{low} is less than δ. This "stop" time is given by

$$t_{high,stop} \equiv \frac{\ln(1/\delta)}{\nu_{min}} \left(\frac{\nu_{min} t_{low,short}}{\ln(1/\delta)} \right)^{T_{low}/T_{high}} \tag{13}$$

where $t_{low,short}$ is the shortest transition time at T_{low}. When this stop time is reached, the system clock is advanced by $t_{low,short}$, and the corresponding transition is accepted. The TAD procedure is then started again in the new basin. Thus, in TAD, two parameters govern the accuracy of the simulation: δ and ν_{min}.

The average boost in TAD can be dramatic when barriers are high and T_{high}/T_{low} is large. However, as TAD relies upon harmonic TST for validity, any anharmonicity error at T_{high} will lead to a corruption of the dynamics. This anharmonicity error can be controlled by choosing a T_{high} that is not too high.

A number of advances have led to increased efficiency in particular systems. "Synthetic" mode [23], a KMC treatment of low-barrier transitions, can significantly improve the efficiency in cases where low-barrier events are repeated often. Furthermore, if we know something about the minimum barrier to leave a given state, either because we have visited the state before and have a lower bound on this minimum barrier or because the minimum barrier is supplied a priori, we can accept a transition and leave the state earlier than the time given by Equation (13) (see ref. 24 for details).

TAD has been demonstrated to be very effective for studying the long-time behavior of defects produced in collision cascades [25,26]. An MD/TAD procedure has also been applied to the simulation of thin-film growth of Ag [27] and Cu [28] on Ag(100). Heteroepitaxial systems are especially hard to treat with techniques such as KMC due to the increased tendency for the system to go off lattice due to mismatch strain, and because the rate catalog needs to be considerably larger when neighboring atoms can have multiple types. Other applications for which TAD has proven effective include defect diffusion on oxide surfaces [29], the diffusion of interstitial clusters in Si [30] and defect diffusion in plutonium [10].

4. RECENT ADVANCES AND APPLICATIONS

While the basics of the three methods described above were established roughly a decade ago, they provide such a fertile ground for further development that they are still the subject of ongoing research. Currently, this research proceeds along three main axes: (i) generalization of the methods to extend their range of applicability, (ii) algorithmic improvements to make the methods more robust and easier to apply, and (iii) creation of hybrid methods by combining AMD methods together or with other simulation approaches. In the following, we describe some recent advances along these three directions and discuss some successful demonstrations and applications.

4.1 Superstate parallel-replica dynamics

While the derivation of the parallel-replica method in Section 1 does not impose a particular definition of a "state" of the system, the operational definition used in practice often corresponds to a single basin of the potential energy surface, i.e., a state is taken to be the ensemble of points of configuration space that converge to the same fixed point under a local minimization of the energy of the system (e.g., using a steepest-descent algorithm). An exponential distribution of escape times is then obtained if the typical timescale for a transition out of the state is long compared to the characteristic vibrational period of the system around that fixed point, i.e., if there is a separation of timescale between vibrations and transitions between basins. While this definition has the virtue of being conceptually and computationally simple, it limits the range of possible applications to systems where the basins are deep enough (relative to $k_B T$) and well separated from each other and leaves many other, more complex, systems out of reach. There is thus a clear need to develop strategies to capitalize on more general definitions of states and hence higher-level gaps in the characteristic timescales spectrum.

For example, in the case of pyrolysis of hexadecane, it was shown that a state could be defined as the ensemble of all configuration space points that share the same network of covalent bonds [9]. In that case, these "superstates" contain a large number of simple energy basins of the potential energy surface, each corresponding to a different conformation of the molecular backbone. There, the method exploited the separation of timescale between the rapid changes of dihedral angles of the backbone (intrasuperstate transitions) and the slow covalent bond breaking process (intersuperstates transitions) rather than between the vibrational timescale and that of sampling of the different dihedral angles. This enables one to ignore the "irrelevant" fast transitions that would demand incessant dephasing and decorrelation and concentrate directly on the real kinetic bottlenecks.

Another common situation that arises is one where the infrequent-event system of interest is coupled to a complex and rapidly evolving environment such that, considered as a whole, the combined system makes transitions so rapidly that any computational gain is impossible using a traditional definition of a state. A specific example of this type of system is a solid surface in contact with a liquid. We might be interested in following the evolution of the surface morphology during electrochemical deposition or etching, or discovering the relevant steps of a surface-catalytic reaction or a corrosive process. In this type of system, a single potential energy basin definition of a state would imply that transitions occur every few femtoseconds. However, the vast majority of these would correspond to local changes in the coordination of liquid atoms, and only very rarely would the configuration of the solid atoms, which are really the variables of interest here, change. Similarly to the case of pyrolysis discussed above, a superstate can be here defined as all the points of configuration space that converge to the same configuration of the slow variables of the system — the position of the atoms belonging to the solid phase — under minimization of the energy.

In order to demonstrate that this definition is proper, we carried out superstate parallel-replica simulations of the diffusion of an adatom on a Ag(100) surface (modeled using the embedded atom method) in contact with a film of a prototypical fluid (modeled using a Lennard-Jones potential). Here the interaction strength of liquid atoms with other liquid atoms or with silver atoms was taken to be 10 times weaker than the corresponding silver–silver interaction. The liquid film was left free to expand to its liquid–vapor equilibrium density. The distribution of 500 adatom hopping times at 600 K is shown in the left panel of Figure 4 for a direct simulation using conventional MD and for a superstate parallel-replica simulation. The results clearly show that transition statistics are equivalent for the two approaches, implying that our superstate definition restores the timescale separation essential for the validity of the method. In this case the simulation was quite efficient, with a parallel efficiency of around 0.8 despite the presence of very fast transitions in the liquid.

This good agreement is not due to a negligible effect of the liquid on the adatom dynamics. Indeed, as shown in the right panel of Figure 4, the hopping kinetics are strongly affected by the liquid. Quantitatively, a fit to a standard Arrhenius rate expression $k(T) = v_0\exp(-\Delta E/k_B T)$, where v_0 is the vibrational prefactor for the transition and ΔE its activation energy, shows that the presence of the liquid significantly increases both ΔE (from 0.53 to 0.70 eV) and v_0 (from 7.17×10^{13} to $6.28 \times 10^{14}\,\text{s}^{-1}$). Interestingly, while the liquid slows down diffusion in the regime we probed, the results suggest that it could actually assist the adatom's diffusion at temperatures exceeding about 900 K. One must however be careful with such extrapolations because the relevant physics could be significantly modified as the liquid approaches its critical point. Indeed these modifications to the kinetics stem from many, often conflicting, factors, and their effects are extremely difficult to obtain accurately from higher-level models. The development of direct but efficient methods to probe the kinetics of complex systems like this is thus extremely useful.

Note that, in the two cases discussed above, chemical intuition was essential in properly defining superstates such that an appropriate separation of timescales was obtained. For more complex systems, intuition alone will not be sufficient and considerable effort might be required to identify an exploitable gap in the characteristic timescales of the system. There is thus a need to develop on-the-fly methods to appropriately define superstates based on MD data alone. Efforts toward this goal are presently under way [31].

4.2 Self-learning hyperdynamics

As illustrated in the previous section, the AMD algorithms often have to be tuned and sometimes even tailored for particular applications. This might be straightforward in some cases, but might require considerable care in others. It would thus be highly desirable if the methods could be made to automatically adapt themselves to every system, or even to every state of every system, in order to deliver optimal performance while maintaining tight control on accuracy.

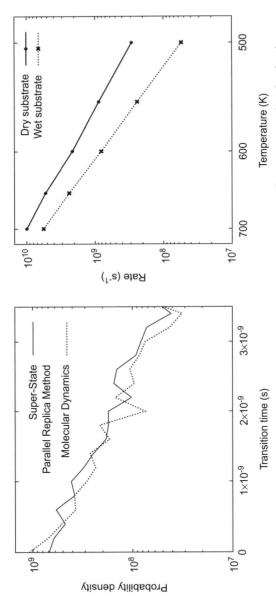

Figure 4 Left: Distribution of hopping times for an adatom at a solid–liquid interface at 600 K for conventional molecular dynamics and for superstate parallel-replica dynamics. Right: Temperature dependence of the hopping rate for an adatom at a dry interface and at a wet interface as obtained using superstate parallel-replica dynamics.

In the current state of affairs, nowhere is this need more pressing than in hyperdynamics.

Indeed, as discussed above, the applicability of hyperdynamics is often hampered by the difficulty in building bias potentials that satisfy all the formal requirements — namely that (i) the bias potential should vanish at any dividing surface between different states and (ii) the kinetics on the biased potential obeys TST — while providing substantial acceleration of the dynamics. Both requirements are very challenging to meet in practice. Indeed, condition (i) must be obeyed for *all* dividing surfaces around *all* states, which, given that the transition pathways or even the possible conformations that the system can adopt are *a priori* unknown, is highly nontrivial to enforce at a reasonable computational cost. An important step in that direction has however been recently taken by Miron and Fichthorn, with the introduction of their "bond-boost" bias potential [32]. As the name suggests, the bond-boost potential is composed of pairwise terms that tend to soften the bonds between atoms. The key assumption here is that transitions between states will involve the formation or breaking of some bond so that the proximity to a transition state will be signaled by an unusually large distortion of a bond. If the overall bias potential is then designed to vanish when *any* bond in the system distorts by more than some critical amount (say by more than 20% of its equilibrium length), then it should be possible to safely turn off the bias before a dividing surface is reached. Now, we will make use of the following bond-boost functional form:

$$\Delta V_b(\varepsilon) = \min_{\alpha}[\delta V_\alpha(\varepsilon_\alpha)] \qquad (14)$$

with $\varepsilon_\alpha = [r_\alpha - r_\alpha^{eq}]/r_\alpha^{eq}$, where r_α and r_α^{eq}, are the current and equilibrium length of bond α, respectively. Following Miron and Fichthorn, we define

$$\delta V_\alpha(\varepsilon_\alpha) = \begin{cases} S_\alpha\left[1 - \left(\dfrac{\varepsilon_\alpha}{q}\right)^2\right], & |\varepsilon_\alpha| \leq q \\ 0, & \text{otherwise} \end{cases} \qquad (15)$$

where q corresponds to the critical bond deformation and S_α controls the strength of the bias applied along bond α. Note that in this formalism, only the bond that has the minimal value of δV will experience a bias force. Of course, the identity of the biased bond will change frequently in the course of the dynamics. Assuming that a conservative value of q is known (which we acknowledge can be a delicate issue), the bond-boost potential can be used to accelerate the dynamics of systems where transitions involve significant bond distortion.

The remaining question is that of the choice of the local bias strengths S_α. Since a suitable choice of q already guarantees that the bias potential vanishes at dividing surfaces, they should be tuned to yield the largest possible boost while making sure that requirement (ii) is respected, namely that TST is obeyed on the biased potential. Typically, this amounts to requiring that the (biased) potential basin corresponding to each state be sampled on a vibrational timescale, or, in other words, that it is free of local minima where the system could get trapped.

The pairwise nature of the bond-boost makes this task easier since such traps would show up as a non-convexity of some of the biased effective pair potentials, which in the canonical ensemble can be taken to be the pairwise potential of mean force (PMF, denoted as $\mathcal{V}$). Thus, assuming that $\mathcal{V}$ is approximately quadratic for $|\varepsilon| < q$, the safety condition can be enforced by setting $S_\alpha \leq \min[\mathcal{V}_\alpha(-q), \mathcal{V}_\alpha(q)]$, so that $\mathcal{V}_\alpha(\varepsilon_\alpha) + \delta V_\alpha(\varepsilon_\alpha)$ is convex over $[-q, q]$.

The pairwise PMF of a system vibrating in state A can be defined as:

$$\mathcal{V}_\alpha^A(r) = -\frac{1}{\beta} \ln\left[\frac{\langle\rho_\alpha(r)\rangle_A}{\langle\rho_\alpha(r_\alpha^{eq})\rangle_A}\right] \tag{16}$$

where $\langle\rho_\alpha(r)\rangle_A$ is the canonical distribution function of the length of bond α. The bias potential can thus be parameterized by computing the relative probability density that each bond is at $\varepsilon = \pm q$ over that of being at $\varepsilon = 0$. A direct-MD evaluation of $\mathcal{V}_\alpha(\pm q)$ would however be prohibitively expensive given that the crossing of any $\varepsilon_\alpha = \pm q$ point is by definition a rare event. Since the set of configurations where any of the $\varepsilon_\alpha = \pm q$ forms a hypersurface at which the bias potential vanishes, hyperdynamics can be used to speed up the evaluation of $\mathcal{V}_\alpha$. The key to avoiding the circular problem where $\mathcal{V}_\alpha^A$ is needed to carry out hyperdynamics but hyperdynamics is needed to efficiently compute $\mathcal{V}_\alpha^A$ is to instead aim at estimating a lower bound. This way, a safe but conservative parameterization of the bias can be turned on even before the convergence of $\mathcal{V}_\alpha^A(q)$ is achieved. As the statistics improve, the lower bound on the PMF can be made tighter and thus the bias potential parameterization more aggressive, until convergence is finally achieved. This feedback loop between improvement of the PMF estimate and acceleration of the rate of improvement of this estimate makes the procedure, termed *self-learning hyperdynamics*, extremely efficient. Once a transition to a new state is detected, the S_α is set to zero, and the procedure is repeated. Under self-learning hyperdynamics, the bias potential is thus continually adjusted to provide optimal performances while ensuring accuracy. The complete mathematical and implementation details of the method will be available in an upcoming publication [33].

The efficiency of the method can be appreciated from Figure 5, where the evolution of hyper-time with MD time under self-learning hyperdynamics is shown for two defects — a silver monomer and trimer — on a Ag(100) surface at room temperature. The results show that, after an initial bootstrapping period during which the bias potential is turned off completely, the learning procedure quickly leads to an increase of the bias strengths, thereby increasing the boost factor, until convergence is finally achieved. One can show that in the early part of the learning process, the hyper-time actually increases exponentially with MD time, which makes the procedure very efficient. This example also clearly shows that the bias potential needs to be tuned for each state if optimal performance is to be achieved. Indeed, for a trimer, the maximal safe boost is found to be around 80 while it exceeds 1,000 for the monomer system. A safe nonadaptive approach would thus demand that the minimal safest boost of *any* of the possible states of the system be used, which, as we demonstrated, may be much smaller than the

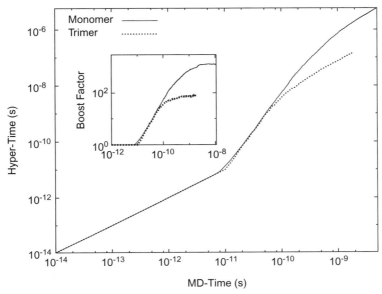

Figure 5 Evolution of the hyper-time as a function of MD time for a monomer and a trimer on a Ag(100) surface at $T = 300$ K under self-learning hyperdynamics. Inset: Evolution of the corresponding boost factors as a function of MD time.

typical safest boost for a given state. Application of this methodology to the study of the mechanical properties of nanowires is currently under way.

4.3 Spatially parallel TAD

While direct-MD simulation with a short-range empirical potential requires $O(N)$ computational work as the system size N is increased, the AMD methods do not exhibit such favorable scaling. This is relatively easy to understand if we consider a system that is made larger in a self-similar way, so that doubling N leads to at least twice as many reactive pathways and twice the total escape rate. A parallel-replica dynamics simulation on the doubled system will reach the first transition in half the wall-clock time, while the dephasing and correlated-event overhead will remain the same. For large enough N, this overhead will dominate the simulation. In hyperdynamics, the bias potential must vanish near every dividing surface, and increasing N increases the fraction of time that the system spends near some dividing surface. Thus, for a valid bias potential, no matter what the form, as N is increased, the boost drops, ultimately approaching unity in the large-N limit.

For TAD, the overall computational work to advance the system by a given time scales at best as $N^{2-\gamma}$, where $\gamma = T_{\text{low}}/T_{\text{high}}$. The power of $-\gamma$ comes from the reduction in $t_{\text{high,stop}}$ from Equation (13) as $t_{\text{low,short}}$ decreases with N, one power of N comes from the cost of each high-T force call, and the other power of N comes from the fact that accepting a transition advances the system by a time inversely

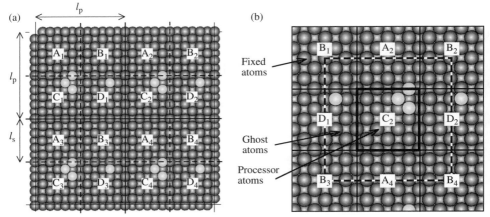

Figure 6 Illustration of the spatial partitioning in the ParTAD method. (a) Lattice and sublattice (SL) partitioning in the SL method on which ParTAD is built. Patches are indicated by numbers, subpatches by letters. (b) Setup for TAD dynamics on one subpatch. Reprinted, with permission, from ref. 34. Copyright 2007 by the American Physical Society. http://link.aps.org/doi/10.1103/PhysRevB.71.125432

proportional to N. If the saddle searches involve all N atoms instead of being localized to a fixed number of atoms, the TAD work at large N is dominated by the cost of these searches, causing an overall scaling of roughly $N^{3+1/3-\gamma}$.

As a consequence of this unfavorable scaling with N, applications of the AMD methods to date have involved at most a few thousand atoms, and have typically been much smaller than that. Since many processes of interest require larger system sizes to capture the essential physics, we are seeking ways to make larger AMD simulations feasible. One important step in this direction is our recent development of a spatially parallel TAD approach [34], which we describe very briefly here.

Spatially parallel TAD, or ParTAD, builds on a parallelization approach originally developed for KMC, the synchronous sublattice (SL) algorithm [35]. In the SL method, the system is spatially divided into a lattice of regions, or patches, each owned by one processor, and each of these regions is further subdivided into SL patches. At any given time, for each of the lattice regions, KMC or TAD dynamics evolve within one, but only one, of the SL patches. After a "cycle time," all processors switch to the next SL patch and again perform dynamics for one cycle time. The simulation then proceeds by repeating this process so that all SL patches cyclically become active.

Figure 6(a) shows an example of a two-dimensional SL definition appropriate for surface diffusion or growth, while Figure 6(b) shows the TAD simulation cell for one processor, consisting of the active SL patch, a "ghost region" of additional moving atoms taken from the adjacent SL patches, and a layer of nonmoving atoms further out. During the TAD dynamics, any attempted event determined to be in the ghost region is excluded from consideration for acceptance, since it is not in the dynamically active patch.

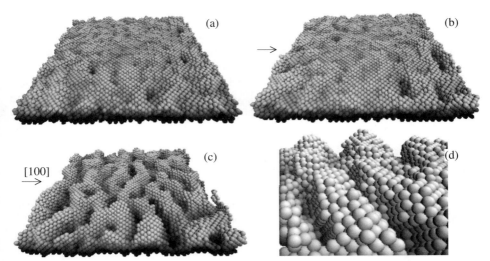

Figure 7   Surface morphology for Cu/Cu(100) films grown to 7-monolayer thickness using ParTAD. (a) Normal-incidence deposition; (b) deposition 30° off normal; (c) deposition 60° off normal; and (d) blow-up of portion of 60° film, showing (100)-oriented cliffs. Arrows indicate the deposition direction. Reprinted, with permission, from ref. 37. Copyright 2008 by the American Physical Society. http://link.aps.org/doi/10.1103/PhysRevLett.101.116101

The important feature in the SL method is that there is always a buffer of dormant SL region separating any two dynamically active regions, which eliminates the difficulty associated with synchronizing events and resolving conflicts at the boundaries between processors. For a finite cycle time, the SL method is not exact, but if the cycle time is comparable to or shorter than a typical reaction time, it is extremely accurate. A detailed discussion and demonstration of the requirements for accuracy in the case of KMC can be found in ref. 35.

Even for vanishing cycle times, the ParTAD method introduces an additional approximation into the dynamics compared to regular TAD; because the subpatch has a fixed size, concerted processes larger than this size, if they exist, will be suppressed. In exchange, the method gives a very favorable computational scaling with system size, which appears to be roughly order $\log(N)$ as processors are added in proportion to N [34].

Metallic film growth is an example where the larger size scale and long-time capability of ParTAD is valuable, since experimental deposition rates are far slower than MD can access, and the morphological features sometimes take on a scale that requires a large simulation cell to prevent artifacts. We have recently studied films of Cu deposited at low temperature onto clean Cu(100) surfaces at varying deposition angles. Our goal was to understand recent X-ray diffraction observations [36] that these films have large surface-normal compressive strain, thought to be due to an enhanced concentration of bulk vacancies. We used MD for the deposition events and ParTAD between deposition events to achieve a growth rate several orders of magnitude slower than we could with MD alone.

Figure 7 shows 7-monolayer (ML) films grown with ParTAD on 36 processors at three different deposition angles at a growth rate of 5,000 ML/s at $T = 40$ K. On this millisecond timescale, activated events do occur, although the experimental growth rate is still much slower (0.02 ML/s).

Our synthesized X-ray diffraction spectrum at a deposition angle of 60° off normal is in very good agreement with experiment, but inspection of the film shows that the large compressive strain arises not from large-scale vacancy incorporation, as had been previously suggested, but rather from nanoscale roughness, as can be seen in Figure 7(c). This roughness, caused by shadowing and the suppression of thermally activated "downward-funneling" events at low temperature, also shows interesting vertical (100) "cliffs" (Figure 7(d)). An analysis of the relevant activation barriers for the suppressed downward-funneling events also leads to an estimate of the critical temperature for the onset of compressive strain $T_c \sim 120$–150 K, in good agreement with experiment.

5. CONCLUSION

Since their introduction a little more than 10 years ago, the AMD methods have proven useful in a variety of situations where the timescales of interest are out of reach of direct MD and where the kinetics are too rich to be adequately described with a limited list of predetermined pathways. When the activation barriers between the different states are high relative to the thermal energy, any of the AMD methods can yield colossal accelerations, providing a view of atomistic dynamics over unprecedented timescales. Further, by leveraging the particular strength of each of the methods, or, as demonstrated above, by generalizing and combining them with other techniques, a wide variety of situations can be efficiently simulated. More discussion of the specific strengths and weaknesses of the methods can be found in a recent review [38].

If the methods have enjoyed considerable successes, they have also sometimes failed to provide significant acceleration. In most, if not all, of the problematic cases, this failure is related to the presence of large numbers of states connected by very low barriers. In this situation, the low barriers limit the boost, and a huge number of transitions may be required before the timescale of interest can be reached. While some strategies have been put forward to mitigate this issue (e.g., superstate parallel-replica dynamics, synthetic TAD [23], state-bridging hyper-dynamics [39]), more work is required before victory can be claimed. For example, an on-the-fly state definition algorithm that automatically identifies an exploitably large separation of timescales would tremendously extend the reach of parallel-replica dynamics, enabling it to address notoriously difficult problems like protein folding, where the energy landscape is extremely rough. Statistical analysis tools could also be used to identify dynamically "irrelevant" states that could be ignored or lumped with others without affecting the long-time dynamics. Many of these ideas are now being explored and will hopefully lead to more general and robust AMD methods in the next few years.

ACKNOWLEDGMENTS

Work at Los Alamos National Laboratory (LANL) was supported by the DOE Office of Basic Energy Sciences and by the LANL Laboratory Directed Research and Development program. LANL is operated by Los Alamos National Security, LLC, for the National Nuclear Security Administration of the US DOE under Contract No. DE-AC52-06NA25396. Work at University of Toledo was supported by NSF grant DMR-0606307.

REFERENCES

1. Eyring, H. The activated complex in chemical reactions. J. Chem. Phys. 1935, 3, 107–15.
2. Pechukas, P. Transition-state theory. Ann. Rev. Phys. Chem. 1981, 32, 159–77.
3. Truhlar, D.G., Garrett, B.C., Klippenstein, S.J. Current status of transition-state theory. J. Phys. Chem. 1996, 100, 12771–800.
4. Voter, A.F. Introduction to the kinetic Monte Carlo method, In Radiation Effects in Solids (eds K.E. Sickafus, E.A. Kotomin and B.P. Uberuaga), Springer, NATO Publishing Unit, Dordrecht, The Netherlands, 2006, pp. 1–24.
5. Voter, A.F. Parallel replica method for dynamics of infrequent events. Phys. Rev. B 1998, 57, 13985–88.
6. Voter, A.F., Montalenti, F., Germann, T.C. Extending the time scale in atomistic simulation of materials. Ann. Rev. Mater. Res. 2002, 32, 321–46.
7. Uberuaga, B.P., Stuart, S.J., Voter, A.F. Parallel replica dynamics for driven systems: derivation and application to strained nanotubes. Phys. Rev. B 2007, 75, 014301-1–9.
8. Uberuaga, B.P., Voter, A.F., Sieber, K.K., Sholl, D.S. Mechanisms and rates of interstitial H_2 diffusion in crystalline C_{60}. Phys. Rev. Lett. 2003, 91(10), 105901–4.
9. Kum, O., Dickson, B.M., Stuart, S.J., Uberuaga, B.P., Voter, A.F. Parallel replica dynamics with a heterogeneous distribution of barriers: application to n-hexadecane pyrolysis. J. Chem. Phys. 2004, 121(20), 9808–19.
10. Uberuaga, B.P., Valone, S.M., Baskes, M.I., Voter, A.F. Accelerated molecular dynamics study of vacancies in Pu. AIP Conf. Proc. 2003, (673)213–5.
11. Uberuaga, B.P., Hoagland, R.G., Voter, A.F., Valone, S.M. Direct transformation of vacancy voids to stacking fault tetrahedra. Phys. Rev. Lett. 2007, 99, 135501-1–4.
12. Mishin, Y., Suzuki, A., Uberuaga, B.P., Voter, A.F. Stick-slip behavior of grain boundaries studied by accelerated molecular dynamics. Phys. Rev. B 2007, 75, 224101-1–7.
13. Duan, Y., Halley, J.W., Curtiss, L., Redfern, P. Mechanisms of lithium transport in amorphous polyethylene oxide. J. Chem. Phys. 2005, 122, 054702-1–8.
14. Warner, D.H., Curtin, W.A., Qu, S. Rate dependence of crack-tip processes predicts twinning trends in f.c.c. metals. Nat. Mater. 2007, 6, 876–81.
15. Shirts, M., Pande, V.S. Screen savers of the world unite! Science 2000, 290(5498), 1903–4.
16. Voter, A.F. A method for accelerating the molecular dynamics simulation of infrequent events. J. Chem. Phys. 1997, 106, 4665–77.
17. Berne, B.J., Ciccotti, G., Coker, D.F. (eds), Classical and Quantum Dynamics in Condensed Phase Simulations, World Scientific, River Edge, NJ, 1998.
18. Valleau, J.P., Whittington, S.G. A guide to Monte Carlo for statistical mechanics: 1. Highways. In Statistical Mechanics. A. Modern Theoretical Chemistry (ed. B.J. Berne), Vol. 5, Plenum, New York, 1977, pp. 137–68.
19. Becker, K.E., Mignogna, M.H., Fichthorn, K.A. Accelerated molecular dynamics of temperature-programmed desorption. Phys. Rev. Lett. 2009, 102, 046101-1–4.
20. Voter, A.F. Hyperdynamics: accelerated molecular dynamics of infrequent events. Phys. Rev. Lett. 1997, 78, 3908–11.
21. Miron, R.A., Fichthorn, K.A. Heteroepitaxial growth of Co/Cu(001): an accelerated molecular dynamics simulation study. Phys. Rev. B 2005, 72, 035415-1–7.

22. Hamelberg, D., Mongan, J., McCammon, J.A. Accelerated molecular dynamics: a promising and efficient simulation method for biomolecules. J. Chem. Phys. 2004, 120, 11919–29.
23. Sørensen, M.R., Voter, A.F. Temperature-accelerated dynamics for simulation of infrequent events. J. Chem. Phys. 2000, 112, 9599–606.
24. Montalenti, F., Voter, A.F. Exploiting past visits or minimum-barrier knowledge to gain further boost in the temperature-accelerated dynamics method. J. Chem. Phys. 2002, 116(12), 4819–28.
25. Uberuaga, B.P., Smith, R., Cleave, A.R., Henkelman, G., Grimes, R.W., Voter, A.F., Sickafus, K.E. Dynamical simulations of radiation damage and defect mobility in MgO. Phys. Rev. B 2005, 71(10), 104102-1.
26. Uberuaga, B.P., Smith, R., Cleave, A.R., Montalenti, F., Henkelman, G., Grimes, R.W., Voter, A.F., Sickafus, K.E. Structure and mobility of defects formed from collision cascades in MgO. Phys. Rev. Lett. 2004, 92(11), 115505-4.
27. Montalenti, F., Sørensen, M.R., Voter, A.F. Closing the gap between experiment and theory: crystal growth by temperature accelerated dynamics. Phys. Rev. Lett. 2001, 87, 126101-1–4.
28. Sprague, J.A., Montalenti, F., Uberuaga, B.P., Kress, J.D., Voter, A.F. Simulation of growth of Cu on Ag(001) at experimental deposition rates. Phys. Rev. B 2002, 66(20), 205415-1–10.
29. Harris, D.J., Lavrentiev, M.Y., Harding, J.H., Allan, N.L., Purton, J.A. Novel exchange mechanisms in the surface diffusion of oxides. J. Phys. Condens. Matter 2004, 16(13), L187–92.
30. Cogoni, M., Uberuaga, B.P., Voter, A.F., Colombo, L. Diffusion of small self-interstitial clusters in silicon: temperature-accelerated tight-binding molecular dynamics simulations. Phys. Rev. B 2005, 71(12), 121203-1–4.
31. Meerbach E., Schütte, C. Sequential change point detection in molecular dynamics trajectories. J. Multivariate Anal. Submitted for publication.
32. Miron, R.A., Fichthorn, K.A. Accelerated molecular dynamics with the bond-boost method. J. Chem. Phys. 2003, 119(12), 6210–6.
33. Perez, D., Voter, A.F. Accelerating atomistic simulations through self-learning bond-boost hyperdynamics. Submitted for publication.
34. Shim, Y., Amar, J.G., Uberuaga, B.P., Voter, A.F. Reaching extended length scales and time scales in atomistic simulations via spatially parallel temperature-accelerated dynamics. Phys. Rev. B 2007, 76, 205439-1–11.
35. Shim, Y., Amar, J.G. Semirigorous synchronous sublattice algorithm for parallel kinetic Monte Carlo simulations of thin film growth. Phys. Rev. B 2005, 71, 1254321-1–14.
36. Botez, C.E., Miceli, P.F., Stephens, P.W. Temperature-dependent vacancy formation during the growth of Cu on Cu(001). Phys. Rev. B 2002, 66, 195413-1–6.
37. Shim, Y., Borovikov, V., Uberuaga, B.P., Voter, A.F., Amar, J.G. Vacancy formation and strain in low-temperature Cu/Cu(100) growth. Phys. Rev. Lett. 2008, 101, 116101-1–4.
38. Uberuaga, B.P., Montalenti, F., Germann, T.C., Voter, A.F. Accelerated molecular dynamics methods. In Handbook of Materials Modeling, Part A: Methods (ed. S. Yip), Springer, Dordrecht, The Netherlands, 2005, p. 629.
39. Miron, R.A., Fichthorn, K.A. Multiple-time scale accelerated molecular dynamics: addressing the small-barrier problem. Phys. Rev. Lett. 2004, 93, 128301-1–4.

Section 3
Bioinformatics

Section Editor: Wei Wang

Department of Chemistry and Biochemistry
University of California at San Diego
La Jolla, CA 92093
USA

CHAPTER 5

Recent Advances on *in silico* ADME Modeling

Junmei Wang[1] and **Tingjun Hou**[2]

Contents

Abstract

Although significant progress has been made on high-throughput screening (HTS) of absorption, distribution, metabolism and excretion (ADME), toxicity, and pharmacokinetic properties in drug discovery, the *in silico* ADME and toxicity (ADME-Tox) prediction still plays an important role in facilitating pharmaceutical companies to select drug candidates wisely prior to expensive clinical trials. Unlike *in vitro* or *in vivo* ADME-Tox assays, *in silico* ADME-Tox is particularly efficient and cheap to search a great number of compounds in screening libraries or virtual molecules in combinatorial chemistry prior to synthesizing them. In the last several years, a lot of new ADME-Tox models have been published and many new software packages and ADME-Tox databases have emerged. In this review, we will present the advances on some oral administration related ADME properties, which

[1] Department of Pharmacology, The University of Texas Southwestern Medical Center, Dallas, TX, USA
[2] Functional Nano & Soft Materials Laboratory (FUNSOM), Soochow University, Suzhou, PR China

Annual Reports in Computational Chemistry, Volume 5
ISSN: 1574-1400, DOI 10.1016/S1574-1400(09)00505-2

include aqueous solubility, Caco-2 and MDCK permeability, blood–brain barrier (BBB), human intestinal absorption (HIA), plasma protein binding (PPB), as well as oral bioavailability (F). We will not only simply review the recent published models but also provide our deep insight on how to construct more accurate and reliable ADME-Tox models.

Keywords: ADME-Tox; solubility; Caco-2 absorption; blood–brain barrier; human intestinal absorption; oral bioavailability; plasma protein binding; QSAR

1. INTRODUCTION

It is assumed that the chemical space exceeds 10^{60} molecules and it is impossible for mankind to make all those molecules. So far, only about 27 million compounds have been registered [1]. However, the number of different small molecules within our own bodies is about a few thousands. As a consequence, it is tough to discover small molecules that actively interact with protein targets since the biological related chemical entities only represent an amazingly small fraction of the entire chemical space. According to the Center for Drug and Evaluation and Research of US Food and Drug Administration (FDA), the numbers of new molecular entities (NME) approved annually, which are shown in Figure 1, did not change significantly over the past two decades. According to Figure 1, evidently there is no trend indicating the number of approved NME is soaring in recent years. Although numerous new technologies have been introduced into the field of drug discovery, such as combinatorial chemistry and high-throughput screening (HTS), in the last 20 years, why is it even more difficult to bring drug candidates into market? Certainly, investigation on the reasons that caused the attrition of drug candidates can help us to discover developable drug candidates prior to expensive clinical trials. In 1991, about 40%

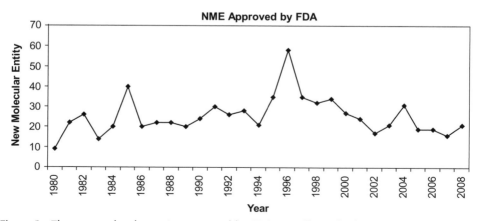

Figure 1 The new molecular entity approved by FDA annually in the last 28 years.

of drug candidate attrition was caused by adverse pharmacokinetics and bioavailability, while by 2000, this factor had dramatically reduced to $\sim 10\%$ and the most dominant reason for attrition was the lack of efficacy, which accounted for $\sim 30\%$ of all attrition [2]. As to attrition caused by drug toxicity, the percentages were about 15% and 25% in 1991 and 2000, respectively. Clearly, the ADME-Tox (absorption, distribution, metabolism, excretion, and toxicity) and pharmacokinetics-related attrition significantly dropped when pharmaceutical industry began to remedy the biggest cause of attrition in 1991. Now HTS of ADME-Tox properties is routinely conducted in pharmaceutical companies. However, the *in vitro* and *in vivo* assays are very time-consuming and costly. Only a tiny fraction of synthesized and screening compounds are selected to do ADME-Tox analysis. *In silico* ADME-Tox modeling, on the other hand, is much more efficient and can deal with large screening libraries. Moreover, *in silico* ADME-Tox models can serve as drug likeness filters to prioritize screening libraries. Those filters typically have better discriminative power than the conventionally used drug likeness filters, like Lipinski's "Rule of Five" [3]. For example, it was found that the real drugs are about 20 times more soluble than those "drug-like" molecules in the ZINC database [4] that pass the "Rule of Five" filter [5]. In the virtual screening study using $\log S$, the logarithm of aqueous solubility in moles per liters, as the filter, less than 50% of "drug-like" molecules in the ZINC database passed the threshold of $\log S$, being larger than -5.0, compared to 85% of real drugs. There is a trend in drug discovery that *in silico* ADME-Tox models are incorporated into the paradigm of drug lead identification and optimization procedures [6–16].

As one of the hot fields in computer-aided drug design (CADD), a number of reviews have been published on the progress of ADME-Tox modeling recently [17–21]. It is not necessary for us to repeat other people's work. Therefore, in this paper we only focus on the latest models published in the last 3 years. ADME-Tox properties can be loosely classified into two categories, namely, the "physicochemical" and "physiological" categories. The physicochemical properties, which include aqueous solubility, logarithm of octanol–water partition coefficient ($\log P$), logarithm of octanol–water distribution coefficient ($\log D$), and pK_a, are governed by simple physicochemical laws. On the other hand, the physiological ADME-Tox properties, which can be further grouped into *in vitro* ADME-Tox properties (such as Caco-2 permeability and MDCK permeability, liver microsomes, etc.) and *in vivo* pharmacokinetic properties (such as oral bioavailability (F), human intestinal absorption (HIA), plasma protein binding (PPB), urinary excretion, area under the plasma concentration–time curve (AUC), total body clearance (Cl), volume of distribution, and elimination half-time ($t_{1/2}$)) are governed by many factors. Take oral bioavailability as an example. Various physiological factors reduce the availability of drugs prior to their entry into the systemic circulation; these factors may include, but are not limited to, poor absorption from the gastrointestinal tract, degradation or metabolism of the drug prior to absorption, and hepatic first-pass effect. In addition, each of these factors may vary from patient to patient and some drugs have multiple circulating active forms. As a consequence, it is rather difficult to build highly predictable models for these properties. The severity is further exaggerated by the lack of high-quality experimental data.

The present paper is organized as follows. First, a brief introduction is given for each ADME-Tox property, especially how it is related to ADME-Tox, and then the latest *in silico* models for that property are discussed. Further, some experts' opinion is presented on how to model that property more accurately and reliably. This is followed by a discussion on how to build up predictable QSAR models with all kinds of statistical tools. Finally, the ADME-Tox resources, including both databases and software packages are summarized.

2. *IN SILICO* MODELING OF KEY ADME-TOX PROPERTIES

2.1 Solubility

Aqueous solubility is one of the major physicochemical properties to be optimized in drug discovery. It is related to "A" and "D" in the ADME-Tox. Aqueous solubility and membrane permeability are the two key factors that affect a drug's oral bioavailability. Because of the importance of aqueous solubility, a lot of effort has been put on developing reliable models to predict this physicochemical property. Although some progress has been made and a lot of models have been constructed, very few models, no matter implemented in commercial packages or in public domain, have satisfactory predictability. In concept, aqueous solubility S of a nonelectrolyte is the concentration (mol/L) of its saturated aqueous solution. Usually, the logarithm of solubility, $\log S$, is used for convenience. Aqueous solubility is almost exclusively dependent on the intermolecular adhesive interactions between solute and solute, solute and water, and water and water. The solubility of a compound is thus affected by many factors that include the size and shape of the molecule, the polarity and hydrophobicity of the molecule, the ability of some groups to participate in intra- and intermolecular hydrogen bonding, as well as the state of the molecule (e.g., additional lattice energy is paid for a compound in the crystalline state to dissolve). One may take these factors into consideration while selecting proper descriptors to build up models and predict this property.

Recently, Goodman et al. sponsored a solubility prediction competition in conjunction with *Journal of Chemical Information and Modeling* to predict the solubilities of 32 molecules [22]. Another 100 drug-like molecules were provided as the training set. The goal of the competition was to find out how well solubility could be predicted with all kinds of means. The *"Solubility Challenge"* is based on intrinsic solubility data measured in one laboratory for a set of biologically relevant compounds. Overall, 99 completed entries were scored and summarized as follows: the number of models was 99; percentage of correct prediction was 41.2% for all 32 compounds and 40.0% if 4 very soluble molecules were excluded ($r^2 = 0.35$). When the four worst-predicted compounds (outliers) were further removed from the 28-compound list, the average percentage of correct prediction increased by 46.7% and average r^2 increased to 0.59. A correct prediction of $\log S$

is defined as the calculated $\log S$ value within $0.5 \log_{10}$ units of experimental data. In our point of view, the overall performance of solubility prediction is not as encouraging as that reported in the published papers. How to construct more reliable models and increase a QSAR model's applicability is discussed later in this review. More performance of solubility models were reviewed by Lipinski et al. [3], Jorgensen et al. [23], Hou et al. [24], and Wang et al. [5,25].

In the following, the latest models published in recent years are briefly discussed. The aqueous solubility models can be categorized into two types: those that correlate with experimentally determined properties and those that do not. The first type of model was exemplified by Jain and Yalkowsky's [26] recent work in which $\log S$ was correlated with experimentally determined melting point (MP) and the logarithm of octanol–water partition coefficient: $n = 580$ and AUE $= 0.45$. Since T_m is not always available, the above equation has a limited application. Wang et al. [25] tried to replace T_m with molecular polarizability, Pol, which can be easily calculated using some empirical models. The correlation coefficient square of T_m to Pol was 0.452 for the clean version of Jain and Yalkowsky's data set ($n = 578$). Encouragingly, the model constructed using calculated $\log P$ ($C \log P$) and Pol was only marginally worse and the standard error and r^2 were 0.887 and 0.905, respectively. The regression equation is shown in the following equation:

$$\log S = 1.095 - 0.008 \times \text{Pol} - 1.078 \times C \log P$$

Again, $\log S$ drops by one logarithm unit if $C \log P$ increases by one or Pol increases by 125. It should be noted that the above empirical rules are very rough and large errors may occur for some unusual molecules. Zhou et al. [27] provided another remedy for the Jain and Yalkowsky's model by calculating MP when it is unavailable experimentally. For the 1,000 organic compounds in the training set, the partial least square (PLS) model achieved a reasonable performance ($r^2 = 0.83$ and RMSE $= 0.85$). The descriptors they used to model MP were SciTegic's extended connectivity fingerprints.

Quite a few second types of models are calculated by summing up each atom type or molecular fragment's contribution in accordance with the additive characteristics of aqueous solubility. No doubt, some molecular-based properties, such as $C \log P$, polar surface area (PSA), and molecular weight (WT), can enter the regression equation pretending they are special "atom types" or "molecular fragments." Klopman and Zhu [28] reported a set of models with the counts of fragments as descriptors, and the best model utilized 171 fragments: $n = 1,168$, $m = 171$, $r^2 = 0.95$, AUE $= 0.49$. In a more recent report by Hou et al. [24], the counts of atom types in addition to two correction factors (hydrophobic carbon and square of molecular weight) were used to build up a model for 1,290 organic molecules that covered a large variety of chemical classes ($n = 1,290$, $m = 78$, $r^2 = 0.92$, AUE $= 0.48$, RMSE $= 0.61$, and $n_{\text{test}} = 120$, AUE$_{\text{test}} = 0.57$, RMSE$_{\text{test}} = 0.79$). Wang et al. [5] reported a set of aqueous solubility models for a larger data set using atom type–classified solvent-accessible surface areas as descriptors. The best model achieved a very encouraging performance ($n = 1,708$, $m = 50$, $r^2 = 0.897$, AUE $= 0.505$, RMSE $= 0.664$, and LOO $q^2 = 0.886$, LOO RMSE$_{\text{test}} = 0.705$).

The fragment/atom type contribution method does not need any descriptors based on other theoretical models; it only needs to count the occurrence of functional groups or atom types in a molecule, so it is extremely time-saving. One potential disadvantage of this kind of method is that new fragments or atom types not defined in the training sets may cause substantial errors.

Besides the counts of fragments or atom types, a lot of theoretical descriptors have been successfully applied in predicting aqueous solubility. Mitchell and Jurs [29] utilized topological, geometric, and electronic descriptors to predict aqueous solubility for a set of diverse molecules ($n = 295$, $m = 9$, $r^2 = 0.93$, RMSE $= 0.64$). Huuskonen [30] developed a set of solubility models for a data set of 1,297 diverse compounds that were described by 24 atom-type E-state indices and 6 topological indices ($n = 884$, $m = 30$, $r^2 = 0.89$, RMSE $= 0.67$ and $n_{test} = 413$, $q^2 = 0.88$, RMSE$_{test} = 0.71$). The Huuskonen data set was also studied by Tetko et al. with purely 38 atom-type E-state indices as descriptors. The best model was generated by artificial neural network (ANN): $m = 33$, $r^2 = 0.91$, and RMSE $= 0.62$ [31]. Again the same data set was used by Liu and So to develop a simple ANN model using 19 descriptors (hydrophobicity, hydrophilicity, molecular weight, and 2D-topolgical indices): $n = 1,033$, $m = 19$, $r^2 = 0.86$, RMSE $= 0.70$, and $q^2 = 0.85$, RMSE $= 0.72$ for leave-one-out cross-validation [32]. In a recent report, Yan and Gasteiger [33] developed two QSPR models for the Huuskonen data set by a multi-linear regression and ANN. The descriptors comprised a set of 32 values of a radial distribution function (RDF) code representing the 3D compounds and 8 additional correction factors that characterize molecular polarizability, relative aromatic and aliphatic degree, and the ability of atoms to participate in hydrogen bonding. A good predictive power was achieved: $n = 797$, $r^2 = 0.79$, AUE $= 0.70$, RMSE $= 0.93$, and $n_{test} = 496$, $q^2 = 0.82$, AUE$_{test} = 0.68$, RMSE$_{test} = 0.79$ for the regression model.

So far all the models introduced above utilized relatively smaller data sets. One potential problem of these models is that the model applicability may be limited, since only a tiny fraction of drug space is represented by a small set of molecules in the training set. Recently, more and more models have been developed using much larger data sets, such as Delaney's model (2,874 compounds) [34], Votano and Parham's model (4,115 compounds) [35], Obrezanova et al.'s model (3,313 compounds) [36], and Wang, Hou, and Xu's model (3,664 compounds) [25].

Delaney studied a much larger data set of 2,874 compounds by using 9 simple descriptors that included calculated log P, molecular weight, aromatic proportion, noncarbon proportion, PSA, etc [34]. The performance of the model was listed as follows: $n = 2,874$, $m = 9$, $r^2 = 0.69$, AUE $= 0.75$, RMSE $= 1.01$. In another report, Votano and Parham [35] constructed a set of models with topological structure indices as descriptors using a variety of data analysis methods. For the data set of 4,115 aromatic compounds, the RMSE of the 772 test set molecules was 0.91, 1.01, and 1.04, for models developed with ANN, PLS, and multiple linear regression (MLR) analysis, respectively. Regarding the 1,874 molecules nonaromatic compounds, the RMSE of the 166 test set molecules was 0.75, 0.87, and 0.88, for the 3 aforementioned analysis methods, respectively.

The disadvantage of this kind of methods is that they are dependent on the descriptors calculated from other theoretical models, and this kind of dependence produces additional difficulties to estimate the solubility of a molecule automatically. Obrezanova et al. [36] successfully applied Gaussian processes (GP) to automatically construct a QSAR model for a large data set, which has 3,313 organic compounds. In total, 108 descriptors, which included 100 two-dimensional SMARTS-based descriptors and 8 molecular properties including $\log P$, molecular weight, topological polar surface area (TPSA), and the McGowan's volume, were calculated. The best GP-derived model achieved $r^2 = 0.85$ and 0.84 for the training set and test set, respectively. However, when this model was applied to predict the solubility of 564 compounds in the Huuskonen's data set, the r^2 dropped to 0.82 and RMSE increased from 0.79 to 0.96 logarithm units.

In 2009, Wang et al. [25] published another set of solubility models for a set of 3,664 compounds. Their best model, ASM-ATC-LOGP, achieved the leave-one-out q^2 and RMSE of 0.832 and 0.840 logarithm units, respectively. In a 10,000 times 85/15 cross-validation test, the model achieved the mean of q^2 and RMSE as 0.832 and 0.841 logarithm units, respectively. Interestingly, when ASM-ATC-LOGP was applied to predict the solubility for compounds in a 1,708-molecule data set, which comprises all 1,297 compounds in the Huuskonen's data set, the AUE and RMSE dropped to 0.54 and 0.72, respectively. This indicates that compounds outside the Huuskonen's data set are more difficult to model.

In summary, most aqueous solubility models have RMSE between 0.7 to 1.0 log units and good performance was relatively easier to achieve for a smaller data set than for larger data sets. Encouragingly, good performance ($RMSE_{test} < 0.85$) was also achieved for the very last models using large data sets, such as Wang et al.'s MLR models. Most solubility models were constructed with linear regression and ANN. ANN usually outperforms linear regression in generating QSPR models, but the difference is much smaller in terms of q^2, AUE, and RMSE for the external test sets. In addition, ANN models may not be interpretable and have a higher chance to be over-fitted than regression.

Dissolvation is a complicated process and it is affected by many physico-chemical factors, such as the ionization state that further depends on the pH, the crystal size of the solute, and the temperature. Unfortunately, only a very limited number of models take these factors into account, mainly because of the limited experimental data under variable conditions. It is expected that better molecular descriptors are developed to account for those factors in the future.

2.2 Caco-2 and MDCK permeability

A drug has to reach its target to cause an effect. Certainly, the obstacles for a drug to reach its target depend on its administrative route. The most natural and generally the safest way to consume a drug is oral. However, an orally administered drug encounters more barriers to reach its target than the ones using other administrative routes, such as intravenously. First of all, an oral drug must be absorbed through the lining of the stomach or small intestine. If it is in solid form, the drug must be dissolved with the help of the stomach's gastric juices. Second, the drug must cross the intestinal epithelial cell barrier to enter the bloodstream. Finally, the free drug in the bloodstream penetrates several layers of lipid membrane to reach its target.

So, what determines the fraction of a drug entering the bloodstream? One of the key factors is the drug's intestinal permeability. A drug can cross the intestinal epithelial cell through two kinds of mechanisms, namely, passive diffusion and active transportation. It is estimated that as much as $\sim 90\%$ of orally marketed drugs possess moderate to high passive membrane permeability irrespective of potential biochemical liabilities, such as P-glycoprotein efflux or first-pass metabolism [37]. In addition to passive diffusion, a small fraction of drugs can be transported by the active transporters, which include both active carrier systems, such as the monocarboxylic acid carrier that transports salicylic acid, and efflux systems, such as P-glycoprotein. In this context, passive membrane permeability can be regarded as an attribute associated with desirable oral absorption, whereas being membrane impermeable may lead to limited oral absorption. In many cases, a threshold value of permeability of 5×10^{-6} cm/s is sufficient to produce near-complete absorption in the absence of any other barriers to absorption.

To meet the need of conducting HTS for ADME-Tox properties, many slow and expensive *in vivo* ADME assays are now being replaced by *in vitro* cell models. For intestinal absorption, Caco-2 cell lines and Madin Darby canine kidney (MDCK) cell lines are widely used to predict the absorption rate of candidate drug compounds across the intestinal epithelial cell barrier. A number of models for Caco-2 cell permeability and MDCK cell permeability have been reported that predict the oral absorption properties of drugs, mostly limited to small organic molecules. Caco-2 and MDCK permeability are related to "A" and "D" in the ADME-Tox.

The *in silico* Caco-2 permeability was predicted by many research groups [37–40]. The following three papers were published recently. Stoner et al. [41] developed two classification models for a large Caco-2 data set, which included 1,195 "active efflux" and 1,823 "'non-active efflux." There were 151 "active influx" in the "non-active efflux" category. The first model was used to differentiate molecules that undergo "active efflux" mechanism from those that undertake passive diffusion mechanism; and the second model was applied to discriminate low-permeability compounds from medium-to-high permeability ones. Both models were constructed by logical regression using 2D-molecular-graph fingerprints GpiDAPH3 as descriptors. The "active influx" molecules were excluded to build the efflux model, and all "active efflux" and "active influx" molecules were excluded to construct the passive permeability classification model. The performance of both models was encouraging: for the efflux model, 67% "active efflux" molecules (21% wrong and 12% no classification) and 88% "passive permeability" molecules (6% wrong and 6% no classification) were correctly predicted. As for the permeability prediction, 48% low-permeability molecules (20% wrong and 30% no classification) and 81% medium- to high-permeability molecules (1.2% wrong and 17.8% no classification) were predicted correctly.

Recently, Jung et al. [42] developed two artificial neural network models to discriminate intestinal barrier-permeable heptapeptides identified by the peroral phage display experiments from randomly generated heptapeptides. There are two kinds of descriptors: one is binary code of amino acid types (each position used 20 bits) and the other, which is called VHSE, is a property descriptor that characterizes the hydrophobic, steric, and electronic properties of 20 coded amino acids. Both types of descriptors produced statistically significant models and the predictive accuracy was about 70%.

More recently, another linear discriminant analysis (LDA) model was constructed for a set of 157 compounds for which P_{Caco-2} was measured [43]. This model, which applied DRAGON descriptors, achieved an accuracy of classification at 91% for the training set and 84% for the test set. When this model was applied to predict a set of 241 drugs for which HIA data were available, good correlation (>81%) was achieved between the two ADME-Tox properties.

In summary, most research groups focus on developing classification models to discriminate low-permeability compounds from medium- to high-permeability ones. The accuracy of correct grouping is ranged from 70% to 90% given the molecules in the data sets undergo the passive permeability mechanism.

2.3 BBB

Blood–brain barrier (BBB) is a complex cellular system whose purpose is to maintain the homeostasis of the central nervous system (CNS) by separating the brain from the systemic blood. The BBB also expresses numerous efflux transporters, such as P-glycoprotein and uptake transporters that can further influence the absorption of a drug in CNS. Understandably, for CNS-active drugs, high penetration of BBB is required; on the other hand, for non-CNS drugs, low penetration is desirable to minimize CNS-related side effects. Generally, lipophilic drugs can cross the BBB by passive diffusion and hydrophilic drugs penetrate the BBB through active transporters. Experimentally, BBB is measured as the ratio of the concentration in the brain to that in the blood. BBB is related to "*A*," "*D*," and "*M*" of ADME-Tox. Now BBB is becoming one of the key ADME-Tox parameters to be optimized in drug discovery. It is usually expressed in logarithm form – log BB.

In the last several years, a set of BBB QSAR models have been developed. One of the top models, developed by Abraham et al. in 2006 [44], reached the predictive limit obtainable from the data set they used. The experimental errors of the log BB measurements were estimated to be ~0.3. Their model utilized linear free energy relationship (LFER) as descriptors. For the 328-molecule data set, r^2 and RMSE of the MLR model were 0.75 and 0.3 log units, respectively. Interestingly, the RMSE for their test set ($n = 164$) was even lower (0.25 log units).

Recently, a hybrid GA/ANN model was developed by Hemmateenejad et al. [45] to predict the log BB of chemicals. There were 123 molecules in the data set.

Electronic descriptors were calculated for the ab initio optimized (RHG/STO-3G) structures. In addition, $\log P$ as a measure of hydrophobicity and different topological indices were also calculated as additional descriptors. A nonlinear model was constructed using ANN with back propagation. Genetic algorithm (GA) was used as a feature selection method. The best ANN model was utilized to predict the log BB of 23 external molecules. The RMSE of the test set was only 0.140 log units, and 98% of variances in the test data were explained.

In 2007, Konovalov et al. presented a set of QSAR models for a large data set [46] that had 328 blood–brain distribution values. Their best model was obtained using a hybrid technique of k-nearest neighbors (kNN) and MLR, that is, the log BB value of a compound was calculated using the MLR model of its neighbors. The standard Euclidian distance was used to measure the pairwise distances $(d_{ii'} = \sum(x_{ij}-x_{i'j})2, j = 1$ to m, m is the number of descriptors). Again, the LFER descriptors were used. The performance of the model was encouraging: the LOO q^2 and RMSE$_{test}$ were 0.766 and 0.290 log units, respectively.

Recently, Wichmann et al. [47] applied several COSMO-RS σ-moments as descriptors to model BBB permeability. The performance of the log BB model was reasonable given only four descriptors were applied: $n = 103$, $r^2 = 0.71$, RMSE $= 0.4$, LOO $q^2 = 0.68$, RMSE$_{test} = 0.42$. The COSMO-RS σ-moments were obtained from quantum chemical calculations using the continuum solvation model COSMO and a subsequent statistical decomposition of the resulting polarization charge densities.

In 2008, Kortagere et al. [48] developed a set of MLR and support vector machine (SVM) models with shape signatures as descriptors for a set of 351 molecules for which experimental BBB data were available. The generalized regression model was tested on 100 external molecules and resulted in correlation $r^2 = 0.65$. The best 2D shape signature models had 10-fold cross-validation prediction accuracy between 80% and 83% and leave-20%-out testing prediction accuracy between 80% and 82%, as well as correctly predicting 84% of BBB+ compounds ($n = 95$) in an external database of drugs. Using a set of in-house molecular descriptors (2D fragments, $\log P$, molecular weight, PSA, and the McGowan's volume), Obrezanova and Gola recently developed a set of GP models for a set of 151 compounds. The best performed GP model ($r^2 = 0.79$, RMSE $= 0.32$, $q^2 = 0.66$, RMSE$_{test} = 0.49$) achieved a significantly better performance than the PLS model ($r^2 = 0.5$, RMSE $= 0.48$, $q^2 = 0.53$, RMSE$_{test} = 0.58$).

In comparison to Caco-2 permeability, more models were built to predict log BB. Most molecule-based descriptors, such as $\log P$, PSA, and molecular electronic properties, were used to construct models with a variety of statistical tools. The best performing models can approach the limit of experimental error, which was estimated to be 0.3 log units.

2.4 Human intestinal absorption (HIA)

Although a drug's intestinal permeability can be estimated through the Caco-2 or MDCK permeability and the binding affinity of the drug interacting with the involved active transporters, it is still not established whether the drug has a high or low oral absorption. This is because oral absorption or HIA is affected by other factors as well, such as the drug's aqueous solubility and metabolic stability. In experiment, HIA is measured by fraction absorption, %HIA, that is, the dose percentage of orally administered drug to reach the hepatic portal vein. Practically, %HIA is defined as the ratio of total mass absorbed divided by the drug dose. As an *in vivo* property, %HIA is related to "A," "D," and "M" of ADME-Tox.

The theoretical prediction of HIA was pioneered by the "Rule of Five" proposed by Lipinski and coworkers [3]. The "Rule of Five" defined several rules for identifying compounds with possible poor absorption and permeability. Lipinski's rule says that, in general, an orally active drug has no more than one violation of the following criteria: no more than 5 hydrogen bond donors (nitrogen or oxygen atoms with one or more hydrogen atoms); no more than 10 hydrogen bond acceptors (nitrogen or oxygen atoms); a molecular weight less than 500 daltons; a $\log P$ value less than 5.0. Note that all numbers are multiples of five, which is the origin of the rule's name. Poor absorption and permeation are more likely to occur when any two of the above-mentioned rules are violated. The disadvantage of the "Rule of Five" is that it can only give a rough classification of molecules, allowing the elimination of only a very limited set of molecules. Now the "Rule of Five" is so popular that most commercial screening libraries only have a small fraction of compounds disobeying this rule. In other words, Lipinski's "Rule of Five" may not be a good rule to prioritize screening libraries.

All kinds of molecular descriptors, including 1D, 2D, and 3D, were used to model the *in vivo* ADME-Tox property. Considering drug molecules must penetrate some membranes to be absorbed, the lipophilicity of a molecule should be a good descriptor. Indeed, $\log P$, $\log D$, PSA, hydrogen bonding ability, and molecular bulkiness were found important in modeling HIA [20,49,50].

In 1997, Palm and coworkers found that an excellent sigmoidal relationship could be established between %HIA and PSA ($r^2 = 0.94$) for a set of 20 drugs covering a wide range of %HIA values in humans. A conclusion was reached that drugs (%HIA > 90%) having a PSA $\leq 61 \text{Å}^2$ were completely absorbed, while drugs having a PSA $\geq 140 \text{Å}$ were absorbed less than 10% [51]. However, this correlation was not true for a much larger data set that had 553 molecules. The correlation coefficient dropped to 0.7 [39]. Grass and Sinko [52] also came out with the similar conclusion that using PSA as a sole predictor cannot predict absorption in humans well.

The logarithm of distribution coefficient, $\log D$, should be a better descriptor than $\log P$ to model membrane permeation, since $\log D$ takes into account the lipophilicity of different ionic forms of a compound in solution. Unfortunately, $\log D$ is not always available experimentally or computationally. Hou and coworkers [49] found that %HIA could correlate to $\log D$ at pH of 6.5 ($r = 0.63$) much better than to $\log P$. Hou and coworkers found that among the 10 descriptors to build up a classification model for %HIA, the TPSA and calculated apparent octanol–water distribution coefficient at pH 6.5 ($\log D_{6.5}$) showed better classification performance than the others. Applying TPSA and $\log D_{6.5}$ together with several other descriptors including number of "Rule of Five" violations and number of hydrogen bond donors, Hou et al. generated a prediction model of %HIA for a set of 435 molecules.

$$\%HIA = 97.12 - 11.48N_{\text{Rule-of-Five}} - 8.99\langle 0.05 - \log D_{6.5}\rangle$$
$$-0.15\langle TPSA - 49.41\rangle + 0.17(\log D_{6.5})^2 + 3.76\langle n_{\text{HBD}} - 7\rangle$$
$$(n = 435,\ r = 0.87,\ RMSE = 12.70,\ F = 277.59)$$

In this equation, the spline terms are denoted with angular brackets. For a spline term $<f(x)-a>$ is equal to zero when $f(x)<a$, or $(f(x)-a)$ otherwise. The regression with splines allows the incorporation of features that do not have a linear effect over their entire range. In this equation, the threshold value a of TPSA is about 49.11 Å, indicating that higher TPSA values lead to low permeation while TPSA lower than 49.11 has no effect on %HIA at all. As to $\log D_{6.5}$, the threshold, a, was found to be 0.05, which means that lower $\log D_{6.5}$ values (<0.05) lead to lower permeation. The spline term of n_{HBD} having a threshold of 7 implied that the number of hydrogen donor larger than 7 causes unfavorable absorption.

Zhao and coworkers [53] also constructed a linear model using the Abraham descriptors. The MLR model possesses good correlation and predictability for external data sets. In this equation, E is an excess molar refraction ($cm^3/mol/10.0$) and S the dipolarity/polarizability, A and B are the hydrogen bond acidity and basicity, respectively, and V is the McGowan characteristic volume ($cm^3/mol/100$). The large coefficients of A and B indicate too polar molecules having poor absorption.

$$\%HIA = 90 + 2.11\,E + 1.70\,S - 20.7\,A - 22.3\,B + 15.0\,V$$
$$(n = 38,\ r^2 = 0.83,\ q^2 = 0.75,\ RMSE = 16\%,\ F = 31)$$

In the following text, a brief review of some recently published HIA models is presented. Recently, Hou and coworkers reported a quite large data set for HIA, which included 647 drug and drug-like molecules collected from a variety of literature sources. Among these 647 molecules, 578 are believed to be transported by passive diffusion. Based on this data set, Hou et al. [49,50] developed a set of prediction models for %HIA.

First of all, a genetic function approximation (GFA) was used to build up the correlation between %HIA and a variety of 1D and 2D molecular descriptors for 455 compounds. The model was able to predict the fractional absorption with an $r = 0.84$ and an AUE of 11.2% for the training set. When it was applied to predict %HIA for a 98-compound test set, it achieved a $q = 0.90$ and $RMSE_{\text{test}} = 7.3\%$. In another work, Hou et al. [49,50] reported two classification models on HIA prediction; one is based on recursive partitioning (RP) technique and the other on SVM. Both models had a good classification performance to discriminate poor HIA ($\%HIA \leq 30\%$) from good HIA ($\%HIA > 30\%$). For the training set, the RP model correctly predicted 95.9% (71/74) of the compounds in poor HIA class and 96.1% (391/407) in good HIA class; while for the test set, all 5 compounds in poor HIA class were correctly predicted and only 3 of 118 compounds in good HIA class were not correctly identified. The SVM model achieved a performance comparable to that of the RP model: for the training set, 97.8% of poor HIA class and 94.5% of the good HIA class 2 were correctly predicted. As for the 123-compound test set, 100% of compounds in poor HIA class and 97.8% in good HIA class were correctly classified.

In 2008, Yan et al. [54] made prediction for a data set of 552 compounds for which HIA experimental data are available. Molecular descriptors were calculated by ADRIANA.Code and Cerius 2 as well. A set of models were constructed with PLS and SVM regression. The best model, which developed with SVM regression, had correlation coefficient of 0.89 and standard error of 16.35%.

In summary, when the HIA data set is clean, that is, all the molecules undergo passive permeability mechanism, HIA can be predicted with a RMSE as low as 12%–16%. The most relevant descriptors to HIA are PSA, $\log D$, $\log P$, etc.

2.5 Oral bioavailability

Oral bioavailability is one of principal pharmacokinetic properties in drug discovery. It represents the percentage of an oral dose that is available to produce pharmacological actions, in other words, the fraction of the oral dose that reaches the system circulation in an active form. By the definition, when a drug is administered intravenously, its bioavailability is 100%. However, when a medication is administered via other routes, especially orally, its bioavailability decreases due to incomplete absorption and first-pass metabolism.

Compared to HIA, the prediction of oral bioavailability of a drug is even more challenging because it is governed by many biological and physicochemical properties, which include the dissolution of the drug in the gastrointestinal tract, the intestinal membrane permeation, the intestinal and hepatic first-pass metabolism, which are not taken into account in HIA, and even the dosage form of the drug. Furthermore, these factors may vary from patient to patient, and even vary in the same patient over time. Whether a drug is taken with or without food will affect absorption, and other drugs taken concurrently may alter absorption and first-pass metabolism. Moreover, disease states affecting liver metabolism or gastrointestinal function will also have an effect.

Bioavailability represents the percentage of an oral dose that is available to produce pharmacological actions. In practice, it is defined as the percentage of oral dose available to the general blood circulation. Bioavailability is related to "*A*" and "*E*" in ADME-Tox. According to the survey of 367 drugs conducted by Wang et al. [55], the average unsigned error (AUE) and root-mean-square error (RMSE) of the experimental measurements are 12.1% and 14.5%, respectively.

Among the several reasons that lead to a true decrease in bioavailability, drug dissolution and gastrointestinal permeability, which control the rate and extent of drug absorption, are the two most important factors. In general, a drug with high solubility and high membrane permeability is considered to be practically exempt from bioavailability problems; a drug having low solubility yet high permeability or high solubility yet low permeability requires careful formulation to improve its dissolution rate; and finally, a drug with poor solubility and permeability is a problematic candidate for administration. In VolSurf [56], an *in silico* ADME-Tox software package implemented in Sybyl6.9 of Tripos Inc. (www.tripos.com), a hybrid parameter (SolyPerm) that combines the scores of the thermodynamic solubility model and the Caco-2 permeation model is applied to classify drug candidates on the basis of both solubility and permeability. The SolyPerm model was constructed using 1,833 training-set molecules with PLS analysis. The more positive the parameter, the greater the chance is for the compound to be bioavailable. Unfortunately, there is no obvious correlation between the experimental bioavailability and the SolyPerm parameter for Wang et al.'s data set [55]. This indicated that more complicated descriptors may be required to build up the relationship.

The pioneers of bioavailability modeling can be traced back to year 2000. Andrews and coworkers [57] developed a regression model to predict bioavailability for 591 compounds. Compared to the Lipinski's "Rule of Five," the false negative predictions were reduced from 5% to 3%, while the false positive predictions decreased from 78% to 53%. The model could achieve a relatively good correlation ($r^2 = 0.71$) for the training set. But when 80/20 cross-validation was applied, the correlation was decreased to $q^2 = 0.58$.

In 2006, Wang and coworkers [55] reported another regression model for the predictions of oral bioavailability using the counts of functional groups as descriptors. A genetic algorithm was employed to find the prediction models with the best combination of functional groups. The final models include 42 functional groups and other 2 molecular descriptors: molecular refractivity and "Rule of Five." The mean r^2 and mean RMSE for the 20 best models were 0.55 and 21.9%, respectively. For the 90/10 cross-validation, the mean r^2 and mean RMSE for the test sets were 0.42 and 24.6%, respectively.

The classification models for predicting oral bioavailability have also been proposed, for example, the model developed by Yoshida and coworkers [58]. The Yoshida's model was developed using a set of physicochemical parameters including distribution coefficients $\log D_{6.5}$ (pH = 6.5) and $\Delta \log D$ ($\log D_{6.5} - \log D_{7.4}$), and 15 functional groups related to well-known metabolic processes. The ordered multicategorical classification method using the simplex technique was applied to assign compounds in the training set to one of four classes. In the leave-one-out cross-validation test, an average of correct classification of 67% was reasonable for bioavailability modeling. The predictive power of the model was evaluated using a separate test set of 40 compounds, of which 60% (95% within the same class) were correctly classified.

Recently, Gasteiger et al. [59] reported several models to predict human oral bioavailability using Hou and Wang's data set. A set of ADRIANA.Code and Cerius2 descriptors were calculated, and MLR analysis was performed. The best linear model had r^2 of 0.18 and RMSD of 31.15. When a set of subsets was cherry-picked so that each subset had either a common functional group or a similar pharmacological activity, the r^2 values were improved and RMSD values dropped. But the performance of those models was still not satisfactory: the standard errors were above 20.0 and r^2 was lower than 0.6.

In 2007, Moda et al. [60] developed a set of Hologram QSAR models for a much smaller data set of 250 molecules in comparison to the above-discussed models. They found that using all atoms, bonds, connections, and chirality to define molecular fragments led to a set of encouraging models and the best one was achieved when the fragment size was from 4 to 7, which had q^2 of 0.7 and r^2 of 0.93. This model was too good to be true since the standard error was only 7.60, much smaller than the average experimental error, which is 14.5 based on 367 experimental data [55].

It is obvious that the available prediction models cannot give reliable estimations for oral bioavailability. The prediction errors were typically larger than 20% except the Moda's model, for which the data set only had 250 molecules; in contrast, all other models utilized a much larger data set.

So, why is bioavailability prediction so challenging? What is the solution to constructing more reliable models? As stated at the beginning, bioavailability is a physiological *in vivo* property and it is governed by many mechanisms. The aqueous solubility, the membrane permeability, as well as the first-pass metabolism all affect a compound's bioavailability. The molecular descriptors utilized by the current models might not account for all the three aspects. Recently, Hou et al. [61] reported an analysis for a database of human oral bioavailability for 768 chemical compounds. The correlations between several important molecular descriptors and human oral bioavailability were investigated and compared with the earlier work reported by Veber [62]. The overall conclusion of Hou et al. analysis is that no molecule-based descriptors (such as $\log P$, PSA) were significantly correlated to bioavailability and it is difficult to construct a reliable model using simple molecular descriptors. Hou et al. also tried to develop a set of rules to discriminate molecules with good oral bioavailability from those having poor bioavailability. However, the performance of the rule-based classification model was not as successful as for HIA.

According to Gasteiger et al. [59], the correlation coefficient r between bioavailability and HIA is 0.498 for 161 compounds. This conclusion inspires us to propose the use of aqueous solubility, descriptors of HIA models, and some rule-based descriptors to predict first-pass metabolism, to model bioavailability. Another research direction for the prediction of oral bioavailability is to develop separate prediction models for different components involved in oral bioavailability, including passive transcellular transport, paracellular transport, carrier-mediated transport, and first-pass metabolism, and then integrate them together. At present, the development of an integrated model is really difficult or even impossible because the predictions for some mechanisms involved in oral bioavailability are really unreliable.

2.6 Plasma protein binding

As discussed above, a drug must cross a set of obstacles to reach its target. The drug is transported to the target organ through the bloodstream. The unbound fraction of the drug crosses again the capillary membrane and diffuses into the fluid of the interstitial space of the tissue. Certainly, the ability of a drug to cross the capillary membrane and penetrate into tissue depends on its physicochemical and physiological properties, especially its membrane permeability, which can be measured by the *in vitro* Caco-2 and MDCK assays. Besides this, human plasma protein binding (PPB) also plays a crucial role in determining the fraction of bioavailable free drug distributed through various tissues. PPB is expressed as the percentage of a drug in the plasma that is bound to plasma proteins at those concentrations of the drug that are achieved clinically. The percentage of PPB may vary from 0% to almost 100%. PPB is an important pharmacokinetic property because only the unbound and nonplasma protein bound fraction of a drug is able to pass the central compartment and diffuse into the interstitial space fluid to reach the target. Drugs with high percent protein binding values tend to have a greater half-life than those with lower values. PPB is related to "D" in ADME-Tox.

Like bioavailability, there are multiple factors that contribute to the experimental errors, which include the analysis and instruments, as well as the experimental conditions, in addition to the disease states that alter the concentration of albumin or other proteins in plasma that bind drugs. According to Wang et al., the AUE and RMSE of the experimental measurements for 266 compounds are 7.8% and 9.9%, respectively. In Volsulf, an *in silico* QSAR model of predicting binding affinity to human serum albumin (HSA) was constructed with PLS using 408 literature compounds in the training set. The cross-validation suggested that 2 significant latent variables (2 components) linearly constructed from 94 Volsulf variables can explain most of variances. Plotting the experimental versus calculated PPB, Wang et al. noticed that there are still a lot of false positives and false negatives, although many high PPB (%PPB > 80%) compounds were correctly predicted.

Recently, Wang et al. [55] applied GA/MLR to build up a set of QSAR models for PPB using the counts of molecular fragments as descriptors. There were 404 high-quality experimental data in the training set. The consensus score of 30 models was found to improve the correlation coefficient and reduce the standard error significantly. Key fragments that may boost or reduce PPB were also identified. For the PPB predicted by the consensus protocol, the r^2 and RMSE were 0.86 and 13.0%, respectively. Considering the experimental error for PPB was about 10%, the performance of this model is encouraging.

Recently, hologram quantitative structure–activity relationship (HQSAR) was conducted by Moda et al. [63] on a series of structurally diverse molecules with known PPB. The best statistical model ($n = 250$, $r^2 = 0.91$, and $q^2 = 0.72$) was used to predict the PPB of 62 test set compounds, and the predicted values were in good agreement with the experimental results ($n_{test} = 62$, $q^2 = 0.86$, $RMSE_{test} = 12\%$). It is indicated that this model used a much smaller data set than the VolSurf and Wang's models.

In 2007, Wichmann et al. [47] applied several COSMO-RS σ-moments as descriptors to model PPB. Unlike the above-mentioned PPB models, which predicted %PPB directly, Wichmann et al. built up a QSAR model to predict human serum albumin binding, $\log K_{HSA}$ (logarithm of %bound/%free), instead. The performance of the $\log K_{HSA}$ model was reasonable given only four descriptors were applied: $n = 94$, $r^2 = 0.67$, RMSE $= 0.33$, LOO $q^2 = 0.63$, $RMSE_{test} = 0.35$. The COSMO-RS σ-moments were obtained from quantum chemical calculations using the continuum solvation model COSMO and a subsequent statistical decomposition of the resulting polarization charge densities. One problem of the model rests in the training set that was too small.

In 2008, Weaver [64] utilized PPB as an example to demonstrate the concept of "domain of applicability" in QSAR researches. The PLS model was constructed using 17 1D and 2D molecular descriptors. The performance of the model was reasonable for such a large data set for PPB modeling ($n = 685$, $q^2 = 0.56$, RMSE $= 0.55$ AUE $= 0.42$, $n_{test} = 210$, $q^2 = 0.58$, $RMSE_{test} = 0.54$, $AUE_{test} = 0.41$). How domain selection protocol affects the prediction performance will be discussed in Section 3.

Generally speaking, PPB is an easier ADME-Tox property to be modeled in comparison to HIA and bioavailability. The prediction errors were only marginally larger than the experimental errors (Wang's model and Moda's model). When PPB is expressed as $\log K_{HSA}$, the RMSEs were about 0.35 to 0.45.

2.7 Drug metabolism stability

When a drug is administrated orally, it undergoes the first-pass metabolism before reaching systematic circulation. Although metabolism occurs as early as the drug is absorbed in the gastrointestinal tract, the primary site of metabolism is the liver. The liver contains the necessary enzymes for metabolism of drugs and other xenobiotics. It is pointed out that although the liver is the primary site for metabolism, virtually all tissue cells have some metabolic activities. All drugs are detoxified and eventually excreted from the body. Drug metabolism is basically a process that introduces hydrophilic functionalities into the drug molecule to facilitate excretion. There are many enzymes involved in this procedure, especially the cytochrome P450 family, which accounts for about 75% of the total metabolism. In drug discovery, drug metabolism stability is a key determinant of several important drug processes *in vivo*, such as drug–drug interaction and drug toxicity. As noted earlier, drug's bioavailability is also affected by this property. A drug's metabolic stability is usually expressed as *in vitro* half-life and intrinsic clearance.

Considering that many enzymes account for drug metabolism stability, it is very difficult to construct a reliable universal QSAR model to predict drug metabolism stability for multiple mechanisms involved. Rather a set of 2D/3D QSAR as well as pharmacophore models were constructed for individual enzymes. Considering many experimental structures are published for the enzymes, molecular docking has also been applied to identify the possible inhibitors that lead to metabolism instability. One tough problem is how to give an overall assessment of a molecule's metabolism stability given all these models are accessible. Further discussion of the advances on *in silico* modeling drug metabolism stability is beyond the scope of this review. Comprehensive reviews on this topic have been presented by Sakiyama et al. [65], Yengi et al. [17],

Chang et al. [66], Li et al. [67], Yamashita et al. [68], Ekins et al. [69], Baranczewski et al. [70], and Singh [71].

3. MODEL CONSTRUCTION

What is the challenge for *in silico* ADME-Tox modeling? In our point of view, the challenge is how to build up an accurate, reliable, and efficient model with good applicability. To answer this question, we may review the procedure of constructing an ADME-Tox model. The basic procedure of QSAR model construction is shown in Figure 2.

To reach this goal, several aspects must be taken into account. First of all, the basis of a good model is sufficient high-quality data. Collecting a number of high-quality data of the ADME-Tox property is not an easy task. Unlike bioinformatics that has most redundant data and toolkits in the public domain than any other informatics, chemical informatics is propagated by the industry and lots of experimental data are not accessible at all. As to the chemical informatics tools for ADME-Tox modeling, most of them are commercial products and only a tiny fraction of these tools are free to everyone. It is estimated that only about 1% of chemical informatics data are in the public domain. This situation makes it a tough task to collect a sufficient number of data to do modeling. Undoubtedly, the lack of extensive and reliable experimental data is an important reason to hinder the development of reliable ADME prediction models. It is particularly true for the *in vivo* oral bioavailability and HIA data, which are usually collected for drugs or drug candidates in clinic trials. In addition, these data may show

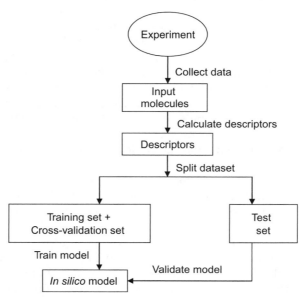

Figure 2 The basic procedure of constructing a QSAR model.

significant variability from one source to another. The big pharmaceutical companies have developed large in-house databases containing consistently measured compound properties. For example, Veber et al. reported an analysis of bioavailability in rats on a data set of over 1,100 compounds studied at GSK-62; Andrews et al. [57] published a study of oral bioavailability for 591 structures from the GlaxoWellcome's internal database. However, these data are usually not available for the scientific community. Many models developed in academia were still based on small historical data sets taken from a variety of literatures, which caused inconsistency between data and lowered the quality of the data sets.

Once the experimental data have been collected, one should make sure that the experimental data are of high quality. Data verification is a tedious task, but it is important. For example, Wang et al. [25] found that there were many erroneous entries with their Bilstein data set for aqueous solubility modeling. The prediction errors of many entries were abnormally large for a set of solubility models. They classified data into three zones: those having residuals of $\log S$ larger than 3.2 or smaller than -3.2 are in the red zone; those within -3.2 to -1.6 or 1.6 to 3.2 are in the yellow zone; and all others are in the green zone. After data verification, they found more than 90% of the entries in red zone and 50% in yellow zone turned out to be erroneous. The most common error arose due to unit transition.

Besides highly reliable experimental data, a high-quality data set for ADME-Tox modeling must have the data points evenly distributed in a broad range, and the molecules of the data set must be diverse to cover more chemical space. The chemical space diversity can be measured by performing pairwise similarity comparison using 2D or 3D fingerprints [64]. The chemical space may be plotted using molecular weight, $\log P$ (or calculated $\log P$), PSA, and the ADME-Tox property to be modeled as axes [5].

In the next step, molecular descriptors are calculated so that the chemical structural information is encoded in molecular descriptors. How to choose molecular descriptors depends on the ADME-Tox property to be modeled. Therefore, some basic knowledge on the ADME-Tox property is needed to choose descriptors wisely. Take HIA as an example. It is a physiological property that is governed by many other procedures such as dissolvation, membrane penetration, etc. Therefore, $\log P$ or $\log D$, which is a measure of lipophilicity, should be a good descriptor. There are thousands of descriptors available, not only for 1D, but also for 2D and 3D. Lots of descriptors are redundant and should be eliminated using covariance analysis. The widely used molecular descriptors include Abraham [44], VolSurf [56], MolSurf [72], Adriana.Code [54], FAF-Drugs2 [73,74], DRAGON [75], Mold [76], Cerius 2 (www.accelrys.com), Molconn-Z (www.tripos.com), and so on.

In the following step, the whole data set is divided into a training set, a cross-validation set, and a prediction set. Lots of modelers randomly divided the whole data set into a training set and a test set and ignored the cross-validation set. We argue that it is not proper to split the whole set randomly into a training set and a test set. In our work of modeling aqueous solubility [25], we performed 10,000 times 85/15 cross-validation test (85% of data randomly selected into training set and 15% into test set) and noticed that q^2, AUE, and RMSE had Gaussian distributions. For one model, ASM-ATC-LOGP, the mean, maximum, and minimum of q^2 were 0.832, 0.884, and 0.762, respectively. The mean, maximum, and minimum of RMSE were 0.841, 0.959, and 0.732, respectively. Obrezanova et al. [36] presented a protocol to do data set splitting using cluster analysis. Compounds in the whole data set were clustered using an unsupervised nonhierarchical clustering algorithm. The chemical structures were represented using 2D path-based fingerprints, and the Tanimoto similarity index was used to measure the distance between two molecules. The algorithm identified dense clusters, where similarity within each cluster was controlled by a user-defined Tanimoto threshold. Compounds that did not belong to any cluster were referred to as singletons. Once the clusters were formed, the cluster centroids and singletons were put into the training set. Finally, for each cluster, the remaining compounds were sorted according to the Y-values and divided into bins. Compounds in each bin were randomly assigned to a training set, a validation set, and a test set. Of course, the probability of compounds entering different sub-data sets was different.

In the next step, models are trained and validated. Many statistical tools have been used to construct QSAR models. The widely used approaches include MLRs, PLS fitting, GAs, ANN, SVMs, GP, and a set of classification algorithms such as RP, k-nearest neighbors, etc. Recently, a new machine-learning approach, GP, which is based on a Bayesian probabilistic approach, has emerged in ADME-Tox modeling [36,77]. GP does not require subjective determination of model parameters and is able to handle a large pool of descriptors and select the important ones. GP is inherently resistant to overtraining and offers a way of estimating uncertainty in predictions. It is a powerful, robust method capable of modeling nonlinear relationships. Typically, the nonlinear approaches, such as ANN, SVM, and GP, produce more accurate models than LMR. However, these nonlinear models are usually difficult to interpret and have a bigger tendency to be over-fitted.

To avoid over-fitting, a commonly used approach is to select a subset of descriptors to build models. GAs are widely used to select descriptors prior to using other statistical tools, such as MLR, to build models. Certainly, principal component analysis and PLS fitting are also widely used in reducing the dimensions of descriptors. Traditionally, stepwise linear regression is used to select certain descriptors to enter the regression equations.

Once a QSAR model is constructed, it must be validated using the external test set. The data points in the test set should not appear in the training set. There are two approaches to improve the prediction accuracy for a given QSAR model. The first approach utilized the concept of "the domain of applicability," which is used to estimate the uncertainty in prediction of a particular molecule based on how similar it is to the compound used to build the model. To make a more accurate prediction for a given molecule in the test set, the structurally similar compounds in the training set are used to construct model and that model is used to make the prediction. In some cases, the domain similarity is measured using molecular descriptor similarity, rather than the structural similarity. The

Table 1 A list of ADME-Tox Software Packages and their major functions

Software Package	Company	LogP	LogD	pKa	Sol	BBB	Caco-2	MDCK	HIA	F	PPB	Met	Tox	P-gp	PA	AP
ACD/labs	ACD labs	✓	✓	✓	✓											
ADME boxes	Pharma Algorithms	✓	✓	✓	✓				✓	✓	✓	✓		✓	✓	
ADME Collection	Scitegic			✓	✓	✓										
Admensa	InPharmatica	✓		✓	✓	✓	✓		✓		✓	✓		✓	✓	
ADMEWORKS	Fujitsu	✓	✓	✓	✓	✓			✓		✓	✓	✓			
CastroPlus	Simulations Plus	✓	✓	✓	✓	✓	✓	✓	✓	✓	✓	✓	✓			✓
Cerius 2	Accelrys	✓			✓	✓			✓			✓	✓			
Compdrug	Compdrug	✓		✓	✓	✓	✓		✓			✓	✓	✓	✓	
iDEA ADME	LION bioscience	✓	✓	✓	✓	✓	✓	✓	✓		✓	✓				
Jchem	ChemAxon	✓	✓	✓						✓						
KnowItAll	Bio-Rad Lab	✓	✓	✓	✓	✓	✓	✓	✓	✓	✓	✓				
Meta	Multicase											✓				
MetaDrug	Genego											✓	✓			
MetaSite	Molecular Discovery											✓				
NorayMet ADME	Noraybio	✓	✓	✓	✓	✓	✓		✓		✓					
PreADMET	PreADMET	✓	✓	✓	✓	✓	✓		✓		✓	✓				
Qikprop	Schrödinger	✓		✓	✓	✓	✓		✓		✓	✓				
StarDrop	Biofocus	✓		✓	✓	✓	✓		✓		✓	✓				
Volsurf	Tripos	✓	✓		✓		✓		✓		✓	✓				

Abbreviations: Sol, aqueous solubility; F, bioavailability; PPB, plasma protein binding; Met, metabolism; Tox, toxicity; PA, passive absorption; AP, apparent permeability.

applications of using domain similarity in ADME-Tox modeling are presented by Konovalov et al. [46] and Weaver et al. [64].

The second possible way to improve the prediction accuracy of the ADME properties is to develop consensus models by combining two or more models for the same property together. Actually, the concept of "consensus model" has been introduced to ADME predictions a long time ago. For example, researchers in Bio-Rad Laboratories, Inc., have introduced the consensus score to the KnowItAll ADME/Tox software system (http://www.bio-rad.com) and found that the employment of multiple complementary models for the same ADME-Tox end point in consensus modeling provides a greater accuracy than that of any single model. Recently, Abshear et al. [78] validated the performance of four models in KnowItAll for predicting intrinsic solubility of 113 diverse organic compounds. For predicted aqueous solubility of 113 compounds, four individual models gave absolute average errors of 0.314, 0.422, 0.327, and 0.324 log units, respectively. By combining these four individual models, the consensus model gave an AUE of 0.257 log units. Wang et al. [55] constructed a set of consensus models to model bioavailability (30 models), PPB (20 models), and urinary excretion (30 models). All these models were constructed by MLR using GA to select descriptors, the count of molecular fragments. More reliable prediction was achieved with the consensus models for all three ADME-Tox properties.

In the last several years, *in silico* ADME-Tox modeling has made substantial progress and numerous QSAR models have been generated. Many databases have been constructed and software packages have been developed. Although it is beyond the scope of this review to introduce these databases and software packages in detail, we have summarized the functions of these ADME-Tox resources and presented in Table 1. We also want to point out that despite the best efforts of their developers many ADME-Tox models including those in commercial packages, which are assumed to be extensively validated, do not have the desired performance. The use of cross-validation between different models is a good way to avoid being misled by poor-quality models in drug discovery.

ABBREVIATIONS

A	absorption
ADME-Tox	absorption, distribution, metabolism, excretion, and toxicity
ANN	artificial neutral network
AUE	average unsigned error
AUE_{test}	average unsigned error of the test set
BBB	blood–brain barrier
D	distribution
E	excretion
F	human oral bioavailability
GA	genetic algorithm
GP	Gaussian processes

HIA	human intestinal absorption
LDM	linear discriminant analysis
$\log D$	logarithm of water distribution coefficient
$\log P$	logarithm of octanol–water partition coefficient
LOO	leave-one-out
M	metabolism
m	number of descriptors
MLR	multiple linear regressions
n	number of data points
n_{test}	number of data points in test set
PLS	partial least square (fitting)
PPB	plasma protein binding
PSA	polar surface area
q^2	cross-validation regression coefficient
QSAR	quantitative structure–activity relationship
QSPR	quantitative structure–property relationship
r^2	regression coefficient
RMSE	root-mean-square error
$RMSE_{test}$	root-mean-square error of the test set
RP	recursive partitioning
SVM	support vector machine
T	toxicity

REFERENCES

1. Dobson, C.M. Chemical space and biology. Nature 2004, 432, 824–8.
2. Kola, I., Landis, J. Can the pharmaceutical industry reduce attrition rates? Nat. Rev. Drug Discov. 2004, 3, 711–5.
3. Lipinski, C.A., Lombardo, F., Dominy, B.W., Feeney, P.J. Experimental and computational approaches to estimate solubility and permeability in drug discovery and development settings. Adv. Drug Deliv. Rev. 1997, 23, 3–25.
4. Irwin, J.J., Shoichet, B.K. ZINC — a free database of commercially available compounds for virtual screening. J. Chem. Inf. Model. 2005, 45, 177–82.
5. Wang, J.M., Krudy, G., Hou, T.J., Zhang, W., Holland, G., Xu, X.J. Development of reliable aqueous solubility models and their application in druglike analysis. J. Chem. Inf. Model. 2007, 47, 1395–404.
6. Zheng, M.Q., Zhang, X.Y., Zhao, M., Chang, H.W., Wang, W., Wang, Y.J., Peng, S.Q. (3S)-N-(L-aminoacyl)-1,2,3,4-tetrahydroisoquinolines, a class of novel antithrombotic agents: synthesis, bioassay, 3D QSAR, and ADME analysis. Bioorg. Med. Chem. 2008, 16, 9574–87.
7. Frecer, V., Berti, F., Benedetti, F., Miertus, S. Design of peptidomimetic inhibitors of aspartic protease of HIV-1 containing -Phe Psi Pro- core and displaying favourable ADME-related properties. J. Mol. Graph. Model. 2008, 27, 376–87.
8. Antunes, J.E., Freitas, M.P., da Cunha, E.F.F., Ramalho, T.C., Rittner, R. In silico prediction of novel phosphodiesterase type-5 inhibitors derived from sildenafil, vardenafil and tadalafil. Bioorg. Med. Chem. 2008, 16, 7599–606.
9. Liao, C.Z., Karki, R.G., Marchand, C., Pommier, Y., Nicklaus, M.C. Virtual screening application of a model of full-length HIV-1 integrase complexed with viral DNA. Bioorg. Med. Chem. Lett. 2007, 17, 5361–5.

10. Deng, J.X., Sanchez, T., Al-Mawsawi, L.Q., Dayam, R., Yunes, R.A., Garofalo, A., Bolger, M.B., Neamati, N. Discovery of structurally diverse HIV-1 integrase inhibitors based on a chalcone pharmacophore. Bioorg. Med. Chem. 2007, 15, 4985–5002.

11. Sivakumar, P.M., Babu, S.K.G., Mukesh, D. QSAR studies on chalcones and flavonoids as anti-tuberculosis agents using genetic function approximation (GFA) method. Chem. Pharm. Bull. 2007, 55, 44–9.

12. Sengupta, D., Verma, D., Naik, P.K. Docking-MM-GB/SA and ADME screening of HIV-1 NNRTI inhibitor: nevirapine and its analogues. In Silico Biol. 2008, 8, 275–89.

13. Sengupta, D., Verma, D., Naik, P.K. Docking mode of delvardine and its analogues into the p66 domain of HIV-1 reverse transcriptase: screening using molecular mechanics-generalized born/surface area and absorption, distribution, metabolism and excretion properties. J. Biosci. 2007, 32, 1307–16.

14. Da Silva, V.B., Andrioli, W.J., Carvalho, I., Taft, C.A., Silva, C. Computer-aided molecular design of novel HMG-CoA reductase inhibitors for the treatment of hypercholesterolemia. J. Theor. Comput. Chem. 2007, 6, 811–21.

15. Ekins, S., Mestres, J., Testa, B. In silico pharmacology for drug discovery: methods for virtual ligand screening and profiling. Br. J. Pharmacol. 2007, 152, 9–20.

16. Ji, H.T., Stanton, B.Z., Igarashi, J., Li, H.Y., Martasek, P., Roman, L.J., Poulos, T.L., Silverman, R.B. Minimal pharmacophoric elements and fragment hopping, an approach directed at molecular diversity and isozyme selectivity. Design of selective neuronal nitric oxide synthase inhibitors. J. Am. Chem. Soc. 2008, 130, 3900–14.

17. Yengi, L.G., Leung, L., Kao, J. The evolving role of drug metabolism in drug discovery and development. Pharm. Res. 2007, 24, 842–58.

18. Wunberg, T., Hendrix, M., Hillisch, A., Lobell, M., Meier, H., Schmeck, C., Wild, H., Hinzen, B. Improving the hit-to-lead process: data-driven assessment of drug-like and lead-like screening hits. Drug Discov. Today 2006, 11, 175–80.

19. Wishart, D.S. Improving early drug discovery through ADME modelling: an overview. Drugs R. D. 2007, 8, 349–62.

20. Hou, T., Wang, J. Structure-ADME relationship: still a long way to go? Expert Opin. Drug Metab. Toxicol. 2008, 4, 759–70.

21. Ekins, S., Waller, C.L., Swaan, P.W., Cruciani, G., Wrighton, S.A., Wikel, J.H. Progress in predicting human ADME parameters in silico. J. Pharmacol. Toxicol. Methods 2000, 44, 251–72.

22. Llinas, A., Glen, R.C., Goodman, J.M. Solubility challenge: can you predict solubilities of 32 molecules using a database of 100 reliable measurements? J. Chem. Inf. Model. 2008, 48, 1289–303.

23. Jorgensen, W.L., Duffy, E.M. Prediction of drug solubility from structure. Adv. Drug Deliv. Rev. 2002, 54, 355–66.

24. Hou, T.J., Xia, K., Zhang, W., Xu, X.J. ADME evaluation in drug discovery. 4. Prediction of aqueous solubility based on atom contribution approach. J. Chem. Inf. Comput. Sci. 2004, 44, 266–75.

25. Wang, J., Hou, T., Xu, X. Aqueous solubility prediction based on weighted atom type counts and solvent accessible surface areas. J. Chem. Inf. Model. 2009, 49, 571–81.

26. Jain, N., Yalkowsky, S.H. Estimation of the aqueous solubility I: application to organic nonelectrolytes. J. Pharm. Sci. 2001, 90, 234–52.

27. Zhou, D.S., Alelyunas, Y., Liu, R.F. Scores of extended connectivity fingerprint as descriptors in QSPR study of melting point and aqueous solubility. J. Chem. Inf. Model. 2008, 48, 981–7.

28. Klopman, G., Zhu, H. Estimation of the aqueous solubility of organic molecules by the group contribution approach. J. Chem. Inf. Comput. Sci. 2001, 41, 439–45.

29. Mitchell, B.E., Jurs, P.C. Prediction of aqueous solubility of organic compounds from molecular structure. J. Chem. Inf. Comput. Sci. 1998, 38, 489–96.

30. Huuskonen, J. Estimation of aqueous solubility for a diverse set of organic compounds based on molecular topology. J. Chem. Inf. Comput. Sci. 2000, 40, 773–7.

31. Tetko, I.V., Tanchuk, V.Y., Kasheva, T.N., Villa, A.E.P. Estimation of aqueous solubility of chemical compounds using E-state indices. J. Chem. Inf. Comput. Sci. 2001, 41, 1488–93.

32. Liu, R.F., So, S.S. Development of quantitative structure-property relationship models for early ADME evaluation in drug discovery. 1. Aqueous solubility. J. Chem. Inf. Comput. Sci. 2001, 41, 1633–9.

33. Yan, A.X., Gasteiger, J. Prediction of aqueous solubility of organic compounds based on a 3D structure representation. J. Chem. Inf. Comput. Sci. 2003, 43, 429–34.
34. Delaney, J.S. Predicting aqueous solubility from structure. Drug Discov. Today 2005, 10, 289–95.
35. Votano, J.R., Parham, M., Hall, L.H., Kier, L.B., Hall, L.M. Prediction of aqueous solubility based on large datasets using several QSPR models utilizing topological structure representation. Chem. Biodivers. 2004, 1, 1829–41.
36. Obrezanova, O., Gola, J.M.R., Champness, E.J., Segall, M.D. Automatic QSAR modeling of ADME properties: blood–brain barrier penetration and aqueous solubility. J. Comput. Aided Mol. Des. 2008, 22, 431–40.
37. Mandagere, A.K., Thompson, T.N., Hwang, K.K. Graphical model for estimating oral bioavailability of drugs in humans and other species from their caco-2 permeability and in vitro liver enzyme metabolic stability rates. J. Med. Chem. 2002, 45, 304–11.
38. Hou, T.J., Zhang, W., Xia, K., Qiao, X.B., Xu, X.J. ADME evaluation in drug discovery. 5. Correlation of Caco-2 permeation with simple molecular properties. J. Chem. Inf. Comput. Sci. 2004, 44, 1585–600.
39. Hou, T.J., Wang, J.M., Zhang, W., Wang, W., Xu, X. Recent advances in computational prediction of drug absorption and permeability in drug discovery. Curr. Med. Chem. 2006, 13, 2653–67.
40. Ma, G.L., Cheng, Y.Y. Predicting Caco-2 permeability using support vector machine and Chemistry Development Kit. J. Pharm. Pharm. Sci. 2006, 9, 210–21.
41. Stoner, C.L., Troutman, M., Gao, H., Johnson, K., Stankovic, C., Brodfuehrer, J., Gifford, E., Chang, M. Moving in silico screening into practice: a minimalist approach to guide permeability screening! Lett. Drug Des. Discov. 2006, 3, 575–81.
42. Jung, E., Kim, J., Kim, M., Jung, D.H., Rhee, H., Shin, J.M., Choi, K., Kang, S.K., Kim, M.K., Yun, C.H., Choi, Y.J., Choi, S.H. Artificial neural network models for prediction of intestinal permeability of oligopeptides. BMC Bioinformatics 2007, 8, 245.
43. Castillo-Garit, J.A., Marrero-Ponce, Y., Torrens, F., Garcia-Domenech, R. Estimation of ADME properties in drug discovery: predicting Caco-2 cell permeability using atom-based stochastic and non-stochastic linear indices. J. Pharm. Sci. 2008, 97, 1946–76.
44. Abraham, M.H., Ibrahim, A., Zhao, Y., Acree, W.E. A data base for partition of volatile organic compounds and drugs from blood/plasma/serum to brain, and an LFER analysis of the data. J. Pharm. Sci. 2006, 95, 2091–100.
45. Hemmateenejad, B., Miri, R., Safarpour, M.A., Mehdipour, A.R. Accurate prediction of the blood–brain partitioning of a large set of solutes using ab initio calculations and genetic neural network modeling. J. Comput. Chem. 2006, 27, 1125–35.
46. Konovalov, D.A., Coomans, D., Deconinck, E., Vander Heyden, Y. Benchmarking of QSAR models for blood–brain barrier permeation. J. Chem. Inf. Model. 2007, 47, 1648–56.
47. Wichmann, K., Diedenhofen, M., Klamt, A. Prediction of blood–brain partitioning and human serum albumin binding based on COSMO-RS sigma-moments. J. Chem. Inf. Model. 2007, 47, 228–33.
48. Kortagere, S., Chekmarev, D., Welsh, W.J., Ekins, S. New predictive models for blood–brain barrier permeability of drug-like molecules. Pharm. Res. 2008, 25, 1836–45.
49. Hou, T.J., Wang, J.M., Li, Y.Y. ADME evaluation in drug discovery. 8. The prediction of human intestinal absorption by a support vector machine. J. Chem. Inf. Model. 2007, 47, 2408–15.
50. Hou, T.J., Wang, J.M., Zhang, W., Xu, X.J. ADME evaluation in drug discovery. 7. Prediction of oral absorption by correlation and classification. J. Chem. Inf. Model. 2007, 47, 208–18.
51. Palm, K., Luthman, K., Ungell, A.L., Strandlund, G., Artursson, P. Correlation of drug absorption with molecular surface properties. J. Pharm. Sci. 1996, 85, 32–9.
52. Grass, G.M., Sinko, P.J. Effect of diverse datasets on the predictive capability of ADME models in drug discovery. Drug Discov. Today 2001, 6, S54–61.
53. Zhao, Y.H., Le, J., Abraham, M.H., Hersey, A., Eddershaw, P.J., Luscombe, C.N., Boutina, D., Beck, G., Sherborne, B., Cooper, I., Platts, J.A. Evaluation of human intestinal absorption data and subsequent derivation of a quantitative structure–activity relationship (QSAR) with the Abraham descriptors. J. Pharm. Sci. 2001, 90, 749–84.

54. Yan, A.X., Wang, Z., Cai, Z.Y. Prediction of human intestinal absorption by GA feature selection and support vector machine regression. Int. J. Mol. Sci. 2008, 9, 1961–76.
55. Wang, J.M., Krudy, G., Xie, X.Q., Wu, C.D., Holland, G. Genetic algorithm-optimized QSPR models for bioavailability, protein binding, and urinary excretion. J. Chem. Inf. Model. 2006, 46, 2674–83.
56. Cruciani, C., Crivori, P., Carrupt, P.A., Testa, B. Molecular fields in quantitative structure-permeation relationships: the VolSurf approach. J. Mol. Struct. Theochem. 2000, 503, 17–30.
57. Andrews, C.W., Bennett, L., Yu, L.X. Predicting human oral bioavailability of a compound: development of a novel quantitative structure-bioavailability relationship. Pharm. Res. 2000, 17, 639–44.
58. Yoshida, F., Topliss, J.G. QSAR model for drug human oral bioavailability. J. Med. Chem. 2000, 43, 2575–85.
59. Wang, Z., Yan, A.X., Yuan, Q.P., Gasteiger, J. Explorations into modeling human oral bioavailability. Eur. J. Med. Chem. 2008, 43, 2442–52.
60. Moda, T.L., Montanari, C.A., Andricopulo, A.D. Hologram QSAR model for the prediction of human oral bioavailability. Bioorg. Med. Chem. 2007, 15, 7738–45.
61. Hou, T.J., Wang, J.M., Zhang, W., Xu, X.J. ADME evaluation in drug discovery. 6. Can oral bioavailability in humans be effectively predicted by simple molecular property-based rules? J. Chem. Inf. Model. 2007, 47, 460–3.
62. Veber, D.F., Johnson, S.R., Cheng, H.Y., Smith, B.R., Ward, K.W., Kopple, K.D. Molecular properties that influence the oral bioavailability of drug candidates. J. Med. Chem. 2002, 45, 2615–23.
63. Moda, T.L., Montanari, C.A., Andricopulo, A.D. In silico prediction of human plasma protein binding using hologram QSAR. Lett. Drug Des. Discov. 2007, 4, 502–9.
64. Weaver, S., Gleeson, N.P. The importance of the domain of applicability in QSAR modeling. J. Mol. Graph. Model. 2008, 26, 1315–26.
65. Sakiyama, Y., Yuki, H., Moriya, T., Hattori, K., Suzuki, M., Shimada, K., Honma, T. Predicting human liver microsomal stability with machine learning techniques. J. Mol. Graph. Model. 2008, 26, 907–15.
66. Chang, C., Ekins, S., Bahadduri, P., Swaan, P.W. Pharmacophore-based discovery of ligands for drug transporters. Adv. Drug Deliv. Rev. 2006, 58, 1431–50.
67. Li, H.Y., Sun, J., Fan, X.W., Sui, X.F., Zhang, L., Wang, Y.J., He, Z.G. Considerations and recent advances in QSAR models for cytochrome P450-mediated drug metabolism prediction. J. Comput. Aided Mol. Des. 2008, 22, 843–55.
68. Yamashita, F., Hara, H., Ito, T., Hashida, M. Novel hierarchical classification and visualization method for multiobjective optimization of drug properties: application to structure–activity relationship analysis of cytochrome P450 metabolism. J. Chem. Inf. Model. 2008, 48, 364–9.
69. Ekins, S., Mestres, J., Testa, B. In silico pharmacology for drug discovery: applications to targets and beyond. Br. J. Pharmacol. 2007, 152, 21–37.
70. Baranczewski, P., Stanczak, A., Sundberg, K., Svensson, R., Wallin, A., Jansson, J., Garberg, P., Postlind, H. Introduction to in vitro estimation of metabolic stability and drug interactions of new chemical entities in drug discovery and development. Pharmacol. Rep. 2006, 58, 453–72.
71. Singh, S.S. Preclinical pharmacokinetics: an approach towards safer and efficacious drugs. Curr. Drug Metab. 2006, 7, 165–82.
72. Norinder, U., Osterberg, T., Artursson, P. Theoretical calculation and prediction of intestinal absorption of drugs in humans using MolSurf parametrization and PLS statistics. Eur. J. Pharm. Sci. 1999, 8, 49–56.
73. Miteva, M.A., Violas, S., Montes, M., Gomez, D., Tuffery, P., Villoutreix, B.O. FAF-Drugs: free ADME/Tox filtering of compound collections. Nucl. Acids Res. 2006, 34, W738–44.
74. Lagorce, D., Sperandio, O., Galons, H., Miteva, M.A., Villoutreix, B.O. FAF-Drugs2: free ADME/Tox filtering tool to assist drug discovery and chemical biology projects. BMC Bioinformatics 2008, 9, 396.
75. Helguera, A.M., Combes, R.D., Gonzalez, M.P., Cordeiro, M. Applications of 2D descriptors in drug design: a DRAGON tale. Curr. Top. Med. Chem. 2008, 8, 1628–55.

76. Hong, H.X., Xie, Q., Ge, W.G., Qian, F., Fang, H., Shi, L.M., Su, Z.Q., Perkins, R., Tong, W.D. Mold(2), molecular descriptors from 2D structures for chemoinformatics and toxicoinformatics. J. Chem. Inf. Model. 2008, 48, 1337–44.
77. Obrezanova, O., Csanyi, G., Gola, J.M.R., Segall, M.D. Gaussian processes: a method for automatic QSAR modeling of ADME properties. J. Chem. Inf. Model. 2007, 47, 1847–57.
78. Abshear, T., Banik, G.M., D'Souza, M.L., Nedwed, K., Peng, C. A model validation and consensus building environment. SAR QSAR Environ. Res. 2006, 17, 311–21.

Section 4
Quantum Chemistry

Section Editor: Gregory S. Tschumper

Department of Chemistry and Biochemistry
University of Mississippi
University, MS 38677
USA

CHAPTER 6

Explicitly Correlated Coupled-Cluster Methods

Toru Shiozaki[1,2], Edward F. Valeev[3] and So Hirata[4]

Abstract

Establishing a hierarchy of rapidly converging, systematic approximations of exact electronic wave functions for general polyatomic molecules is the holy grail of electronic structure theory. Explicitly correlated coupled-cluster (CC-R12) methods, which have recently been developed by us up to a high rank and are reviewed in this chapter, form such a hierarchy; the CC-R12 energies converge most rapidly toward the exact solutions of the Schrödinger equations of stable molecules with respect to both the cluster excitation rank and the one-electron basis-set size. The R12 methods in this

[1] Quantum Theory Project, Department of Chemistry, University of Florida, Gainesville, FL, USA
[2] Department of Applied Chemistry, Graduate School of Engineering, The University of Tokyo, Tokyo, Japan
[3] Department of Chemistry, Virginia Institute of Technology, Blacksburg, VA, USA
[4] Quantum Theory Project and The Center for Macromolecular Science and Engineering, Department of Chemistry and Department of Physics, University of Florida, Gainesville, FL, USA

Annual Reports in Computational Chemistry, Volume 5
ISSN: 1574-1400, DOI 10.1016/S1574-1400(09)00506-4

review are meant to encompass the so-called F12 methods, the term often invoked to distinguish the methods with nonlinear correlation functions (F12) from the linear one (R12).

Keywords: explicit-r_{12} correlation; coupled-cluster methods; higher-order excitations

1. INTRODUCTION

Establishing a hierarchy of rapidly converging, generally applicable, systematic approximations of exact electronic wave functions is the holy grail of electronic structure theory [1]. The basis of these approximations is the Hartree–Fock (HF) method, which defines a simple noncorrelated reference wave function consisting of a single Slater determinant (an antisymmetrized product of orbitals). To introduce electron correlation into the description, the wave function is expanded as a combination of the reference and excited Slater determinants obtained by promotion of one, two, or more electrons into vacant virtual orbitals. The approximate wave functions thus defined are characterized by the manner of the expansion (linear, nonlinear), the maximum excitation rank, and by the size of one-electron basis used to represent the orbitals.

Rapid convergence of molecular properties with respect to the excitation rank is obtained with the systematic approximations of coupled-cluster (CC) theory [2,3]. The CC methods represent a wave function using a nonlinear (exponential) expansion, which ensures size-extensivity and, thereby, uniform high accuracy for molecules of all sizes, including solids [4].

The convergence of molecular properties with respect to the one-electron basis-set size is more problematic. Absolute molecular energies and, by extension, relevant energy differences, pose a particularly severe test for the basis-set convergence. Thanks to the development of systematic series of basis sets [5–7], the convergence of the molecular energy is usually monotonic and smooth, but always slow. Hence, the complete-basis-set limits cannot be reached in practice even with the largest basis sets and some types of empirical complete-basis-set extrapolations [8–10] are necessary. The slow basis-set convergence of the energy is ascribed to the qualitatively incorrect behavior of Slater determinants [11,12] near electron–electron coalescence points, i.e., where two electrons approach each other closely. At these points, the interelectronic repulsion operator (r_{ij}^{-1}) is singular, causing cusps in the exact wave function [13–15]. Slater determinants cannot have such electron–electron cusps because they are constructed from one-electron orbitals only. A more rational and accurate wave function expansion should have explicit interelectronic distance (r_{ij}) dependence to reproduce the correct shape in the vicinity of the singularities of the interelectronic potential.

Hylleraas was first to use a wave function with explicit r_{ij}-dependence in his 1929 breakthrough study of the helium atom [16,17]. Applications of these so-called *explicitly correlated* methods [18] to molecular systems were difficult owing to the need to evaluate expensive and numerous nine- and higher-dimensional integrals. To overcome these difficulties, Kutzelnigg proposed to use the resolution of the identity (RI) in an atomic partial-wave basis to reduce the higher-dimensional integrals [19]. This technical step is the foundation of the so-called R12 method [20–22]. The original R12 method [23] required rather large basis sets, but its applications at the explicitly correlated second-order Møller–Plesset perturbation (MP2-R12) level were possible for systems as large as ferrocene [24] and benzene dimer [25]. Vigorous developments since 2002 have culminated into the modern R12 method, characterized by the use of a separate (auxiliary) basis set for RI [26–28], nonlinear r_{ij}-dependence of wave functions [29], and other technical developments [30]. Since then the R12 method has emerged as the *de facto* standard explicitly correlated approach for use in chemistry owing to the following favorable features: (1) it brings a dramatic reduction in the basis-set errors, requiring only a moderate triple-ζ basis set to recover the correlation energies obtained with the conventional MP2 method with a much larger quintuple-ζ basis set; (2) it can be combined with all existing standard electron-correlation models; (3) the extra cost of R12 terms is relatively small since it only needs integrals with the same dimensionality as those found in the standard methods; and (4) it is completely general and can be applied to polyatomic molecules and extended systems.

A combination of the R12 method with the CC theory is therefore especially desirable since the R12 approach can correct the gravest weakness of the standard CC methods, the large basis-set errors. Noga et al. [31,32] were first to implement the CC-R12 method. Their achievement was a landmark study in electronic structure theory because it allowed molecular energies to be computed for many small systems more accurately than ever before. Because their implementation was not based on the modern R12 technology, large basis sets were required for such computations and, therefore, relatively small systems could be studied. The purpose of this article is to review the considerable advances in the CC-R12 methods that occurred since the pioneering work of Noga et al. and since the review on the R12 methods by one of the authors [20] appeared in this journal 3 years ago. The emphasis will be placed on the implementations of the *complete* CC-R12 method including through and up to connected quadruple excitation operator, i.e., CCSD-R12, CCSDT-R12, and CCSDTQ-R12, reported by the present authors [33–35]. The term "complete" is used in the sense that we take into account *all* numerous diagrammatic terms that arise in the modern formulation of these methods; the implementation of Noga et al. skipped the most complicated of such terms. The R12 methods in this review are meant to encompass the so-called F12 methods, the term often invoked to distinguish the methods with nonlinear correlation functions (F12) from the linear one (R12). The CC-R12 methods developed by us can, therefore, use nonlinear correlation functions such as the Slater-type correlation function of Ten-no [29], providing the descriptions of wave functions that converge extremely rapidly toward the exact solutions of the Schrödinger equations with respect to both excitation rank and basis-set size. They hold great promise as a reliable computational tool for gas-phase thermochemistry, reaction dynamics, structural chemistry, and intermolecular interactions as well as perhaps even solid-state physics [4,36] by virtue of their vastly improved cost-accuracy performance.

2. R12 METHOD FOR A TWO-ELECTRON SYSTEM

To expose the essence of the R12 method of Kutzelnigg [19], consider the simplest two-electron system, the helium atom in its ground state. The exact wave function in the vicinity of an electron–electron coalescence point $\mathbf{r}$ can be expressed [13] as

$$\Psi(\mathbf{r}_1, \mathbf{r}_2) = \left(1 + \frac{1}{2}r_{12}\right)\Psi(\mathbf{r}, \mathbf{r}) + \mathcal{O}(r_{12}^2) \tag{1}$$

or, equivalently,

$$\left.\frac{\partial\Psi(\mathbf{r}_1, \mathbf{r}_2)}{\partial r_{12}}\right|_{\mathbf{r}_1=\mathbf{r}_2=\mathbf{r}} = \frac{1}{2}\Psi(\mathbf{r}, \mathbf{r}) + \mathcal{O}(r_{12}) \tag{2}$$

While any approximate wave function Φ that consists of Slater determinants violates this cusp condition, the product $(1 + (1/2)r_{12})\Phi$ may satisfy it exactly. Inspired by this knowledge and by the fact that a reference wave function (Φ_0) still dominates in the exact correlated wave function, Kutzelnigg has suggested [19] correcting only the short r_{12}-behavior of the reference wave function in the framework of the determinant-based correlation methods:

$$\Psi(\mathbf{r}_1, \mathbf{r}_2) = [1 + f(r_{12})]\Phi_0(\mathbf{r}_1, \mathbf{r}_2) + \chi(\mathbf{r}_1, \mathbf{r}_2) \tag{3}$$

where $f(r_{12})$ is a *correlation factor*, e.g., a linear factor $f(r_{12}) = cr_{12}$, with an adjustable parameter c, and χ denotes a linear combination of excited Slater determinants. Although this wave function cannot satisfy the cusp condition at every point,

$$\left.\frac{\partial\Psi(\mathbf{r}_1, \mathbf{r}_2)}{\partial r_{12}}\right|_{\mathbf{r}_1=\mathbf{r}_2=\mathbf{r}} = \left.\frac{df(r_{12})}{dr_{12}}\right|_{r_{12}=0}\Phi_0(\mathbf{r}, \mathbf{r}) \neq \frac{1}{2}\Psi(\mathbf{r}, \mathbf{r}) \tag{4}$$

it can satisfy it in an average sense. Furthermore, the ansatz in equation (3) allows difficult multidimensional integrals to be avoided via efficient RI methods (see below).

Thus, one can think of the R12 wave function as representing many-body correlation effects through two types of terms: $f(r_{12})\Phi_0$ responsible for the short-range two-body Coulomb correlation and χ describing conventional many-body correlation. The second term is expanded in terms of Slater determinants composed of orbitals from a finite orbital basis set (OBS):

$$\chi(\mathbf{r}_1, \mathbf{r}_2) = \sum_{p<q} c_{pq}|pq\rangle \tag{5}$$

where p and q are spin orbitals expanded by the OBS, $|pq\rangle$ is the excited Slater determinant formed by p and q, and amplitudes c_{pq} are determined from the amplitude equations of some kind. The explicitly correlated term captures the remainder of the correlation energy from excited determinants in the complete-basis set that are not collected by the second term that is based on a finite basis set. Therefore, the first term can be *formally* written in terms of the complete-basis-set orbitals κ and λ as

$$f(r_{12})\Phi_0(\mathbf{r}_1, \mathbf{r}_2) = \sum_{k<\lambda} |k\lambda\rangle \langle k\lambda|f(r_{12})|\Phi_0\rangle \tag{6}$$

However, in practice, the reliance on the complete (or an extremely large) basis set is avoided by exploiting the analytical cancellation of the Coulomb singularities by the correlation factor in evaluating the necessary integrals of the Hamiltonian:

$$\sum_{k<\lambda} \langle pq|r_{12}^{-1}|k\lambda\rangle \langle k\lambda|f(r_{12})|\Phi_0\rangle = \langle pq|r_{12}^{-1}f(r_{12})|\Phi_0\rangle \tag{7}$$

With an appropriate $f(r_{12})$ function, e.g., in the original *linear* form $f(r_{12}) = r_{12}$, the operator product $r_{12}^{-1}f(r_{12})$ is no longer singular. Such cancellation is not possible with Slater determinants alone and this is what allows explicitly correlated wave functions to achieve accurate correlation energies with relatively small basis sets. With the single explicitly correlated term, therefore, we effectively include a linear combination of an infinite set of Slater determinants, but without the need to solve an infinite set of equations to determine the corresponding amplitudes. The R12 method constructs wave functions that are more compact and computationally tractable than naive Slater-determinant-based counterparts.

In the following sections, we describe how this approach can be applied to many-electron systems in the context of the MP2 and CC methods. In the interest of developing theories that are independent of the number of particles, the language of second quantization must be used. Nevertheless, the essential ideas of this section will apply throughout.

3. MP2-R12

3.1 Formalisms

It is instructive to discuss the MP2-R12 method [37] before going into more involved CC-R12. As in MP2, the wave function of MP2-R12 ($|\Psi\rangle$) is a linear combination of the reference HF determinant ($|\Phi_0\rangle$) and doubly excited determinants produced by the action of a two-electron excitation operator $\hat{T}$:

$$|\Psi\rangle = (1 + \hat{T})|\Phi_0\rangle \tag{8}$$

The $\hat{T}$ operator in MP2-R12,

$$\hat{T} = \hat{T}_2 + \hat{\mathcal{J}}_2 \tag{9}$$

includes the standard two-electron excitation operator of the MP2 method,

$$\hat{T}_2 = \frac{1}{4} t_{ij}^{ab} \{a^\dagger b^\dagger ji\} \tag{10}$$

as well as the *geminal excitation* operator,

$$\hat{\mathcal{J}}_2 = \frac{1}{8} \hat{Q}_{12} F_{xy}^{\alpha\beta} t_{ij}^{xy} \{\alpha^\dagger \beta^\dagger ji\} \tag{11}$$

where $F_{xy}^{\alpha\beta}$ is a two-electron integral over a correlation function $f(r_{12})$, t_{ij}^{ab}, and t_{ij}^{xy} are unknown amplitudes to be determined by the procedure described below, and the summation is implied over all repeated indices.[1] The action of operator $\hat{\mathcal{J}}_2$ replaces pairs of occupied orbitals ij with geminal pairs, i.e., what essentially corresponds to orbital products xy multiplied by the correlation factor $f(r_{12})$.[2] In other words, $\hat{\mathcal{J}}_2$ is formally a two-electron excitation operator that promotes electrons into the virtual space represented by the complete-basis set. This operator is parameterized by a relatively small number of unknown coefficients (t_{ij}^{xy}) to take care of short-range two-electron correlations more efficiently than $\hat{T}_2$. The role of the orthogonality projector $\hat{Q}_{12}$ is to ensure that the geminal pairs are orthogonal to the usual products of virtual orbitals produced by $\hat{T}_2$.

The unknown amplitudes (t-amplitudes) in the $\hat{T}$ operator are determined by minimization of the Hylleraas functional [40]:

$$E_{\text{MP2-R12}} = \langle \Phi_0 | \hat{T}^\dagger \hat{f} \hat{T} | \Phi_0 \rangle + 2 \langle \Phi_0 | \hat{T}^\dagger \hat{v} | \Phi_0 \rangle \tag{12}$$

where $\hat{f}$ is the Fock operator and $\hat{v}$ the two-electron part of the Hamiltonian in the normal-ordered second quantization, $\hat{H} = E_{\text{HF}} + \hat{f} + \hat{v}$. By differentiating this functional with respect to the amplitudes, we obtain the MP2-R12 amplitude equations that are written as follows:

$$\langle \Phi_{ij}^{ab} | \hat{v} + \hat{f}\hat{T} | \Phi_0 \rangle = 0 \tag{13}$$

$$\langle \Phi_{ij}^{kl} | \hat{v} + \hat{f}\hat{T} | \Phi_0 \rangle = 0 \tag{14}$$

where the so-called geminal displacement configuration is defined by $|\Phi_{ij}^{kl}\rangle = (1/2) F_{kl}^{\alpha\beta} \{\alpha^\dagger \beta^\dagger ji\} | \Phi_0 \rangle$. Inserting the solution of these linear equations into equation (12), the following simple expression for the MP2-R12 energy is obtained:

$$E_{\text{MP2-R12}} = \langle \Phi_0 | \hat{v}\hat{T} | \Phi_0 \rangle \tag{15}$$

Equations (13) and (15) reduce to the usual MP2 correlation energy expression by substituting $t_{ij}^{xy} = 0$ and, therefore, MP2 and MP2-R12 as defined above converge to the same complete-basis-set limit, however, at vastly different rates. These

---

[1]Here, i and j label occupied spin orbitals, a and b virtual (unoccupied) spin orbitals within the OBS space, and α and β are virtual spin orbitals expanded by the complete basis set. The set of virtual orbitals a is included in the set of virtual orbitals α.

[2]There are different choices of the geminal-generating orbitals x and y: the $ijpq$[38,39] and $ijkl$ ansatz. We employ the $ijkl$ ansatz, which defines x and y as the occupied spin orbitals.

equations are also obtainable by retaining only the lowest-order terms in the corresponding equations of the CCSD-R12 method.

3.2 Algorithms

While the formalisms of MP2-R12 are by themselves simple, designing robust and efficient algorithms is not necessarily so. This is mainly because the high-rank many-electron integrals (up to four-electron integrals) arise as a result of the introduction of the correlation factor in the formalisms and they are extremely expensive to evaluate. For instance, the brute-force evaluation of four-electron integrals needs at least $O(n^8)$ arithmetic operations and is not feasible. In the key molecular integrals, the Coulomb singularities as well as the complete-basis-set indices (α, β) are eliminated by the analytical cancellation with the correlation factor, but in other numerous, less important integrals, the complete-basis-set indices remain. What makes the R12 method practical is the ingenious scheme that eliminates such expensive computational steps and the complete-basis-set indices with the aid of the RI approximation [19]. The central idea is that, by replacing the identity $1 = \sum_{\alpha=1}^{\infty} |\phi_\alpha(1)\rangle\langle\phi_\alpha(1)|$ by an approximation $1 \approx \sum_{p=1}^{n_{RI}} |\phi_p(1)\rangle\langle\phi_p(1)|$ (where $\{\phi_p\}$ is the RI basis set consisting of n_{RI} auxiliary functions), a high-rank many-electron integral can be decomposed into a sum of products of two-electron integrals that are far less expensive to evaluate. A four-electron integral appearing in the MP2-R12 equations can thus be approximated as

$$\langle ijkl|f(r_{12})r_{14}^{-1}f(r_{34})|lkji\rangle \approx \sum_{p,q} \langle ij|f(r_{12})|pj\rangle\langle pl|r_{12}^{-1}|lq\rangle\langle kq|f(r_{12})|ji\rangle \tag{16}$$

The right-hand side can be evaluated with mere $O(n_{occ}^4 n_{RI}^2)$ arithmetic operations (n_{occ} is the number of occupied spin orbitals) using just computationally tractable two-electron integrals. For atoms, nonzero contributions to this sum occur only from the RI basis functions with angular quantum numbers up to $3L_{occ}$, where L_{occ} is the maximum angular quantum number of occupied spin orbitals.

In the original formulation of the R12 method [23], which is often referred to as "the standard approximation" or SA, a large uncontracted OBS was needed as the RI basis, strongly limiting the applicability of the R12 method to relatively small systems. Klopper and Samson [26] later pioneered the use of a separate RI basis set and considerably improved the applicability and accuracy of MP2-R12. Valeev developed an alternative formulation of the RI procedure [27] known as the complementary auxiliary basis set (CABS) approach, which had smaller RI errors. Ten-no, on the other hand, proposed the use of a numerical quadrature for the same purpose [28]. Kedžuch et al. later used the CABS approach to develop a particularly elegant and practical formalism for the MP2-R12 energies [30].

Another critical advance responsible for the success of MP2-R12 is the introduction of nonlinear correlation functions, in particular, the Slater-type correlation function, $1-\exp(-\gamma r_{12})$, of Ten-no [29]. It is asymptotically linear in r_{12} near the coalescence point and, hence, satisfies the cusp condition in the leading order. Unlike the linear correlation function, the Slater-type correlation function exhibits a more desirable long-range behavior and thus should be more appropriate for applications to extended systems. It is also believed to be nearly optimal among single-parameter correlation functions [41] and closely matching in performance multiparameter correlation functions [42]. Although all necessary integrals over Slater-type correlation functions can be evaluated analytically [43], it is often expanded as a linear combination of Gaussian functions [41].

The MP2-R12 methods in their modern formulation have been implemented in several software packages [44–48]. The MP2-R12 energies can now be routinely computed for general polyatomic molecules, providing conventional MP2 correlation energies of quintuple-ζ quality with just a triple-ζ computational cost [41] (see also refs. 49, 50 for numerical examples). The applications of MP2-R12 to large molecules can greatly benefit from the density fitting and local approximations by Manby, Ten-no, and Werner [51–54] as well as from the development of parallel computer algorithms [55].

4. CC-R12

4.1 Formalisms

The ansatz of CC-R12 was introduced by Noga et al. [31,32] as a generalization of the MP2-R12 ansatz to CC theory. Taking CC-R12 with connected singles and doubles (CCSD-R12) as an example, its wave function can be parameterized as

$$|\Psi\rangle = \exp(\hat{T})|\Phi_0\rangle \tag{17}$$

$$\hat{T} = \hat{T}_1 + \hat{T}_2 + \hat{\mathcal{J}}_2 \tag{18}$$

with

$$\hat{T}_1 = t_i^a \{a^\dagger i\} \tag{19}$$

where t_i^a is a single excitation amplitude. The t-amplitudes (t_i^a and t_{ij}^{ab}) and geminal amplitudes (t_{ij}^{kl}) are determined by insisting that the wave function of equation (17) satisfy the Schrödinger equation in the space of singly and doubly excited configurations from the reference as well as in the space of the geminal displacement configurations. Defining the CCSD-R12 similarity-transformed Hamiltonian by $\bar{H} = [\hat{H} \exp(\hat{T})]_C$, where "C" indicates that the operators in the brackets must be diagrammatically connected, the amplitude equations of CCSD-R12 can be written as

$$\langle \Phi_i^a | \bar{H} | \Phi_0 \rangle = 0 \tag{20}$$

$$\langle \Phi_{ij}^{ab} | \bar{H} | \Phi_0 \rangle = 0 \tag{21}$$

$$\langle \Phi_{ij}^{kl} | \bar{H} | \Phi_0 \rangle = 0 \tag{22}$$

The corresponding correlation energy $E_{\text{CCSD-R12}}$ is given by

$$E_{\text{CCSD-R12}} = \langle \Phi_0 | \bar{H} | \Phi_0 \rangle \tag{23}$$

Equations (20), (21), and (23) are analogous to those that define the CCSD method [3] and the R12 ansatz adds another matrix equation (equation (22)) whose rank is equal to that of t_{ij}^{kl}. This ansatz was put forth by Noga et al. [31,32], who also evaluated the left-hand sides of these equations for CC-R12 with connected singles, doubles, and triples (CCSDT-R12) and reported the result in diagrammatic expressions [32], which was far more complex than the corresponding result of CCSDT. Note that, although the above ansatz of Noga et al. is most widely used and also natural in the sense that the contribution of double excitations dominates in dynamical electron-correlation energies, other ansatz are also conceivable such as the ones that include unconventional one- [56] and/or three-electron basis functions.

4.2 Approximate CC-R12 methods and their algorithms

Owing to its complexity, the CC-R12 method was initially realized in various approximate forms. The first implementation of the CCSD-R12 method including noniterative connected triples [CCSD(T)-R12] was reported by Noga et al. [31,32,57–60] within the SA. The use of the same basis set for the orbital expansion and the RI in the SA rendered many diagrammatic terms to vanish and, thereby, drastically simplified the CCSD-R12 amplitude equations, easing its implementation effort. However, the simplified equations also meant that large basis sets (such as uncontracted quintuple-ζ basis set) were needed to obtain reliable results and, therefore, the SA CCSD-R12 method was useful only in limited circumstances.

Subsequently, Klopper and coworkers developed the CCSD(R12) and CCSD(T)(R12) methods [61–63] in which the use of the SA was avoided, while maintaining the simplicity of the equations. The "(R12)" approximation retains the terms that are at most linear in t_{ij}^{kl} and thus simplifies the amplitude equations considerably. Equations (20)–(22) are, therefore, replaced by [61]

$$\langle \Phi_i^a |[\hat{H}(1 + \hat{\mathcal{J}}_2)\exp(\hat{T}_1 + \hat{T}_2)]_C|\Phi_0\rangle = 0 \tag{24}$$

$$\langle \Phi_{ij}^{ab} |[\hat{H}(1 + \hat{\mathcal{J}}_2)\exp(\hat{T}_1 + \hat{T}_2)]_C|\Phi_0\rangle = 0 \tag{25}$$

$$\langle \Phi_{ij}^{kl} |[\hat{f}(1 + \hat{\mathcal{J}}_2)\exp(\hat{T}_1 + \hat{T}_2) + \hat{v}\exp(\hat{T}_1 + \hat{T}_2)]_C|\Phi_0\rangle = 0 \tag{26}$$

This approximation can be justified from a perturbation theory viewpoint that assumes the smallness of t_{ij}^{kl} and is analogous to the treatment of connected triples in CCSDT-1 [64]. The simplification in the equations allowed the CCSD(R12) and CCSD(T)(R12) methods to be implemented by a modest extension of the computational elements developed in the MP2-R12 implementations. Since they do not rely on the SA, they need an auxiliary basis set for the RI, but the rapid basis-set convergence can be obtained.

Valeev and coworkers further advanced the perturbation theory argument, allowing t_{ij}^{kl} to be determined noniteratively. This method, termed CCSD(2)$_{\overline{R12}}$

[65–67], computes the noniterative R12 correction (i.e., the residual basis-set correlation energy) using the following expression [66]:

$$E_{(2)_{\overline{R12}}} = \langle\Phi_0|\hat{\mathcal{J}}_2^\dagger \hat{f}\hat{\mathcal{J}}_2|\Phi_0\rangle + 2\langle\Phi_0|\hat{\mathcal{J}}_2^\dagger \hat{v}|\Phi_0\rangle \qquad (27)$$

where the geminal amplitudes are determined by solving the amplitude equations

$$\langle\Phi_{ij}^{kl}|\hat{v} + \hat{v}\hat{T}_2 + \hat{f}\hat{\mathcal{J}}_2|\Phi_0\rangle = 0 \qquad (28)$$

in which $\hat{T}_2$ operator was obtained by solving the conventional CCSD equations. Note the close resemblance between the energy and geminal amplitude equations of the MP2-R12 method (equations (12) and (14)) and their CCSD(2)$_{\overline{R12}}$ counterparts (equations (27) and (28)). All of the quantities in these equations are necessary in the MP2-R12 and CCSD implementations and hence CCSD(2)$_{\overline{R12}}$ can be relatively easily implemented by combining existing MP2-R12 and CCSD codes. Another technical advantage of the perturbative approach is that no modification of the CCSD program is necessary.

An even more radical yet effective approximation to the R12 method was proposed by Ten-no [28,43], in which the coefficients multiplying the correlation function were held fixed at the values implied by the first-order cusp condition and hence were not to be determined iteratively or noniteratively. Several variants of the CCSD(T)-R12 methods were developed on the basis of this promising approximation by Adler et al. [68], Tew et al. [69], Bokhan et al. [70], and Torheyden et al. [66].

4.3 Full CC-R12 methods and their algorithms

The developments of the aforementioned approximate CC-R12 methods preceded those of the full CC-R12 methods based on nontruncated algebraic equations. The performance and validity of the former must be assessed in comparison with the benchmark results obtainable from the latter. However, until recently the full CC-R12 methods have been intractable owing to the immense complexity associated with the derivation of their working equations and efficient computer implementations. The ansatz of a CC-R12 method of a given rank involves one more set of equations (the so-called geminal amplitude equation) than that of the conventional CC method of the same rank. Each t-amplitude equation in CC-R12, e.g., the t_2-amplitude equation (21), has much more diagrammatic terms than the corresponding equation in CC. The transformation of these working equations into programmable computational sequences also involves many more symbolic algebra steps in CC-R12 than in CC: the identification and isolation of the special R12 intermediates, the RI insertion with the CABS to facilitate the high-dimensional molecular integral evaluations, etc.

The initial development of the full CCSD-R12 method and its higher-ranked analogues, CCSDT-R12 and CCSDTQ-R12, reported recently by us [33–35], was predicated on the following two advances. First, the key technical enhancements mentioned in the introduction had been established and the process of deriving the working equations of the R12 extension of an existing electron-correlation theory and implementing them was, at least conceptually, well understood. Second, the time-consuming and error-prone process of the derivation and implementation of high-rank, non-R12 CC methods had been automated completely by computer algebra [71–73]. Combining these two advances, the implementation of the full CC-R12 methods up to a high rank has been achieved for the first time.

Several important extensions had to be made to computer algebra for CC-R12. It is crucial for the implemented programs to exploit the index-permutation symmetry of not just the molecular integrals and t-amplitudes but also of the numerous intermediate quantities that are products of the integrals and amplitudes. The symmetry ultimately reflects the antisymmetry of wave functions of many fermions and its neglect will make the programs too slow and consume too much memory. In non-R12 methods, the index-permutation symmetry of intermediates is known *a priori* for a range of methods that have been of general interest [72,74] and this knowledge has been used to design the computer algebra. The intermediates occurring in CC-R12, however, have much more diverse symmetry characteristics and the same strategy cannot be employed. The computer algebra SMITH developed by us for CC-R12, instead, determines the index-permutation symmetry of each intermediate and its optimal usage on a case-by-case basis using the following general symmetry rules discovered by us: (i) only indices of the same kind (covariant or contravariant; hole, particle, or CABS) can be permutable; (ii) the indices that originate from the same input tensor are permutable; (iii) the indices that are contracted with those of one input tensor are permutable; (iv) the external indices are permutable. SMITH also performs additional symbolic manipulation and optimization steps necessary for the derivations and implementations of CC-R12: strength reduction (matrix chain multiplication), factorization, common sub-expression elimination, loop fusion, RI insertion with CABS, and special intermediate isolation. The details can be found in refs. 33–35. Note that a similar scheme was subsequently explored by Köhn et al. [75], which was based on the string-based, general-order CC method of Kállay et al. [74,76,77].

Figure 1 illustrates an efficient computational sequence for evaluating the left-hand side of the CCSD-R12 geminal amplitude equation, equation (22), suggested by SMITH. The length and complexity of the equations in this figure attest to the need for computer algebra such as SMITH to derive them correctly and expediently. Unlike the computational sequences of the CC t-amplitude equations, those of the CC-R12 geminal amplitude equations have multiple hotspots (time-consuming matrix multiplication steps) and the significance of computer-assisted algorithm optimization seems greater. For instance, Figure 1 has numerous intermediates (Ξ's) that are rather expensive to compute and should, therefore, be precomputed, stored, and reused (in contrast, the common subexpression elimination was rather unimportant in CC [71]). One of these reusable intermediates $\tilde{t}$ is the product Ft and appears frequently throughout the CC-R12 equations. Identifying these products and treating each product as a

$$(\Xi_0)_{i_7 a'_8}^{j_3 i_4} = +t_{i_7}^{a_9} F_{a'_8 a_7}^{j_3 i_4 *}$$

$$(\Xi_1)_{i_2 a_7}^{i_8 a'_6} = +t_{i_2}^{a_9} v_{a_7 a_9}^{i_8 a'_6}$$

$$(\Xi_2)_{a_6}^{d'_7} = +t_{i_9}^{a_8} v_{a_6 a_8}^{i_9 d'_7}$$

$$(\Xi_3)_{i_2 d'_7}^{i_8 d'_5} = +t_{i_9}^{a_5} v_{d'_7 a_8}^{i_9 d'_5}$$

$$(\Xi_4)_{i_2 d'_7}^{i_8 d'_5} = +t_{i_9}^{a_5} v_{d'_7 a_9}^{i_9 d'_5}$$

$$(\Xi_5)_{d_8}^{i_7} = +t_{i_{10}}^{a_9} v_{a_8 a_9}^{i_7 i_{10}}$$

$$(\Xi_6)_{i_2 a_8}^{i_7 i_9} = +t_{i_{10}}^{a_9} v_{a_8 a_9}^{i_7 i_9}$$

$$(\Xi_7)_{i_2 a_7}^{i_8 d'_6} = +t_{i_6 a_9}^{d'_6 a_9} v_{a_6 a_7}^{i_8 i_{10}}$$

$$(\Xi_8)_{a_6}^{d'_7} = +\tfrac{1}{2} t_{i_9 i_{10}}^{a_7 a_8} v_{a_6 a_8}^{i_9 i_{10}}$$

$$(\Xi_9)_{a_6}^{d'_5} = +\tfrac{1}{2} t_{i_9 i_{10}}^{a_7 a_8} v_{d'_6 a_8}^{i_9 i_{10}}$$

$$(\Xi_{10})_{a'_7}^{d'_5} = +\tfrac{1}{2} t_{i_9 i_{10}}^{a_5 a_8} v_{d'_7 a_9}^{i_9 i_{10}}$$

$$(\Xi_{11})_{a'_7}^{d'_5} = +\tfrac{1}{2} t_{i_9 i_{10}}^{a_5 a_8} v_{d'_7 a_8}^{i_9 i_{10}}$$

$$(\xi_{8,2})_{a'_7}^{d'_6} = +P_2 P_2 t_{i_1 i_8}^{a'_5 d'_7} (\xi_{8,0})_{i_2 a'_7}^{i_8 d'_6} + P_2 \tilde{t}_{i_1 i_8}^{a'_5 a_7} (\xi_{8,1})_{i_2 a_7}^{i_8 d'_6} + P_2 \tilde{t}_{i_1 i_2}^{a'_5 d'_4} (\xi_{8,2})_{d'_7}^{a'_6}$$

$$(\xi_7)_{i_5 i_6}^{j_3 i_4} = +\tfrac{1}{2} t_{i_5 i_6}^{i_7 i_8} X_{i_7 i_8}^{j_3 i_4}$$

$$(\xi_{8,1})_{i_2 a_7}^{i_8 d'_6} = -(V^\dagger)_{i_2 a_5}^{i_6 d'_7} v_{i_2 a_5}^{j_3 i_4} + \tfrac{1}{2} t_{i_2}^{a_6} (V^\dagger)_{a_5 a_6}^{j_3 i_4}$$

$$(\xi_{8,0})_{i_2 d'_4}^{a'_5 i_6} = +P_2 P_2 t_{i_1 i_8}^{a'_5 d'_7} (\xi_{8,1})_{i_2 a_7}^{i_8 d'_6} + P_2 \tilde{t}_{i_1 i_8}^{a'_5 a_7} (\xi_{8,1})_{i_2 a_7}^{i_8 d'_6} + P_2 \tilde{t}_{i_1 i_2}^{a'_5 d'_4} (\xi_{8,2})_{d'_7}^{a'_6}$$

$$(\xi_{4,1})_{a_6}^{d'_7} = -f_{a_6}^{a_7} (\Xi_0)_{i_7 d'_5}^{j_3 i_4} + F_{a_5 d'_7}^{j_3 i_4 *} (\xi_{4,1})_{a_6}^{d'_7}$$

$$(\xi_4)_{a_5 d'_6}^{j_3 i_4} = -f_{a_6}^{a_7} (\Xi_0)_{i_7 d'_5}^{j_3 i_4} + F_{a_5 d'_7}^{j_3 i_4 *} (\xi_{4,1})_{a_6}^{d'_7}$$

$$(\xi_{3,7})_{a'_7}^{d'_6} = -t_{i_9}^{a_8} v_{a_7 a_8}^{i_9 a_6} - \tfrac{1}{2} t_{i_9 i_{10}}^{a_7 a_8} v_{a_7 a_8}^{i_9 i_{10}}$$

$$(\xi_{3,6})_{a'_7}^{d'_5} = -(\Xi_3)_{a'_7}^{d'_5} - (\Xi_{10})_{a'_7}^{d'_5}$$

$$(\xi_{3,5})_{a'_7}^{d'_6} = -t_{i_9}^{a_8} v_{a_7 a_8}^{i_9 a_6} + \tfrac{1}{2} t_{i_2 i_{10}}^{a_6 a_8} v_{a_7 a_8}^{i_9 i_{10}}$$

$$(\xi_{3,4})_{i_9}^{i_8 a_6} = -v_{i_2 a_7}^{i_8 a_6} + f_{i_9}^{a_9} v_{a_7 a_9}^{i_8 i_{10}} + t_{i_2 i_{10}}^{a_6 a_8} v_{d_9 a_7}^{i_8 i_{10}}$$

$$(\xi_{3,3})_{i_2 d'_7}^{i_8 d'_5} = -v_{i_2 d'_7}^{i_8 d'_5} + (\Xi_4)_{i_2 d'_7}^{i_8 d'_5}$$

$$(\xi_{3,2})_{i_2 d'_7}^{i_8 d'_6} = -v_{i_2 a_7}^{i_8 d'_6} + f_{i_2}^{a_9} v_{d_7 a_9}^{i_8 d'_6} - \tilde{t}_{i_2 i_{10}}^{d'_6 a_9} v_{d'_7 a_9}^{i_8 i_{10}}$$

$$(\xi_{3,1})_{i_2 a_7}^{i_8 d'_5} = +(\Xi_1)_{i_2 a_7}^{i_8 d'_5} - (\Xi_1)_{i_2 a_7}^{i_8 d'_5} + (\Xi_7)_{i_2 a_7}^{i_8 d'_5}$$

$$(\xi_{3,0,3})_{i_2 a_8}^{i_7 i_9} = -v_{i_2 a_8}^{i_7 i_9} + (\Xi_6)_{i_2 a_8}^{i_7 i_9}$$

$$(\xi_{3,0,2})_{i_2 d'_4}^{i_7 i_9} = -v_{i_2 a_8}^{i_7 i_9} + t_{i_2}^{a_{10}} v_{d_8 a_9}^{i_7 i_9}$$

$$(\xi_{3,0,0})^{i_7}_{d'_8} = -f_{a_8}^{a_9} - t_{i_{10}}^{a_9} v_{d_8 a_9}^{i_7 i_{10}}$$

$$(\xi_{3,0})_{i_1 i_2}^{i_7 d'_5} = +t_{i_1 i_9}^{a_5 a_8} (\xi_{3,0,0})_{d'_8}^{i_7} + \tfrac{1}{2} P_2 t_{i_1}^{a_8} (\Xi_1)_{i_2 a_8}^{i_7 d'_5} + P_2 \tilde{t}_{i_1 i_9}^{a_5 a_8} (\xi_{3,0,2})_{i_2 a_8}^{i_7 i_9}$$
$$+ P_2 \tilde{t}_{i_1 i_9}^{a_8 a_8} (\xi_{3,0,3})_{i_2 a_8}^{i_7 i_9} - \tilde{t}_{i_1 i_2}^{a_5 a_8} (\Xi_5)_{a_8}^{i_7}$$

$$(\xi_3)_{i_1 i_2}^{i_7 d'_5} = +t_{i_7}^{a_6} (\xi_{3,0})_{i_1 i_2}^{i_7 d'_5} + P_2 \tilde{t}_{i_1 i_8}^{a_6 a_7} (\xi_{3,1})_{i_2 a_7}^{i_8 d'_5} + P_2 \tilde{t}_{i_1 i_8}^{a_5 a_7} (\xi_{3,2})_{i_2 a_7}^{i_8 d'_6}$$
$$+ P_2 \tilde{t}_{i_1 i_8}^{a_5 a_6} (\xi_{3,3})_{i_2 d'_7}^{i_8 d'_5} + P_2 \tilde{t}_{i_1 i_8}^{a_5 a_7} (\xi_{3,4})_{i_2 a_7}^{i_8 a_6} + \tilde{t}_{i_1 i_2}^{a_5 a_7} (\xi_{3,5})_{a_7}^{a_6}$$
$$+ \tilde{t}_{i_1 i_2}^{a_5 a_6} (\xi_{3,6})_{d'_7}^{d'_5} + \tilde{t}_{i_1 i_2}^{a_5 a_7} (\xi_{3,7})_{a_7}^{a_6}$$

$$(\xi_{2,1})_{i_5 i_6}^{i_7 i_8} = +\tfrac{1}{2} t_{i_9 i_{10}}^{i_7 i_8} V_{i_5 i_6}^{i_9 i_{10}}$$

$$(\xi_2)_{i_5 i_6}^{j_3 i_4} = +B_{i_5 i_6}^{j_3 i_4} + P_{i_5 i_6}^{j_3 i_4} + V_{i_7 a_8}^{j_7 a_6} (\Xi_0)_{i_7 d'_6}^{j_3 i_4} + \tfrac{1}{2} X_{i_7 i_8}^{j_3 i_4} (\xi_{2,1})_{i_5 i_6}^{i_7 i_8}$$

$$(\xi_{1,1,0})_{i_2 a_9}^{i_7 i_8} = -v_{i_2 a_9}^{i_7 i_8} + \tfrac{1}{2} t_{i_1 i_2}^{a_{10}} v_{a_9 a_{10}}^{i_7 i_8}$$

$$(\xi_{1,1})_{i_1 i_2}^{i_7 a'_6} = +P_2 t_{i_1}^{a_9} (\xi_{1,1,0})_{i_2 a_9}^{i_7 i_8} + \tfrac{1}{2} t_{i_1 i_2}^{a_9 a_{10}} v_{a_9 a_{10}}^{i_7 i_8}$$

$$(\xi_{1,0,0})_{d'_8}^{i_7} = -f_{a_8}^{a_7} - (\Xi_5)_{d_8}^{i_7}$$

$$(\xi_{1,0})_{i_1 i_2}^{i_7 a'_5} = -f_{i_2}^{a_7} + t_{i_1}^{a_8} (\xi_{1,0,0})_{a_8}^{i_7} - \tfrac{1}{2} t_{i_2}^{a_8 a_9} v_{d_8 a_9}^{j_7 i_{10}} - \tfrac{1}{2} t_{i_2 i_{10}}^{a_8 a_9} v_{a_8 a_9}^{i_7 i_{10}}$$

$$(\xi_1)_{i_1 i_2}^{i_5 i_6} = +P_2 t_{i_1 i_7}^{i_5 i_6} (\xi_{1,0})_{i_1 i_2}^{i_7 a'_5} + \tfrac{1}{2} t_{i_1 i_8}^{i_5 i_6} (\xi_{1,1})_{i_1 i_2}^{i_7 i_8}$$

$$(\xi_{0,0})_{a_5 a_6}^{d'_7} = -f_{a_6}^{a_7} - (\Xi_2)_{a_6}^{d'_7} - (\Xi_8)_{a_6}^{d'_7} + (\Xi_9)_{a_6}^{d'_5}$$

$$(\xi_0)_{a_5 a_6}^{j_3 i_4} = +(V^\dagger)_{a_5 a_6}^{j_3 i_4} + P_2 F_{a_5 a_6}^{j_3 i_4 *} (\xi_{0,0})_{d'_6}^{d'_7} + v_{a_5 a_6}^{j_3 i_4} (\Xi_0)_{i_7 a'_6}^{j_3 i_4}$$

$$\delta_{i_1 i_2}^{j_3 i_4} = +(V^\dagger)_{i_1 i_2}^{j_3 i_4} + \tfrac{1}{2} t_{i_1 i_2}^{a_5 a_6} (\zeta_0)_{a_5 a_6}^{j_3 i_4} + \tfrac{1}{2} X_{i_5 i_6}^{j_3 i_4} (\zeta_1)_{i_5 i_6}^{j_3 i_4} + \tfrac{1}{2} t_{i_1 i_2}^{i_5 i_6} (\zeta_2)_{i_5 i_6}^{j_3 i_4}$$
$$+ F_{a'_5 a_6}^{j_3 i_4 *} (\zeta_3)_{i_1 i_2}^{a_5 a_6} + \tilde{t}_{i_1 i_2}^{i_5 a_6} (\zeta_4)_{a_5 a_6}^{j_3 i_4} + v_{i_1 i_2}^{i_5 a_6} (\Xi_0)_{i_5 d'_6}^{j_3 i_4} + P_2 t_{i_1}^{a_5} (\zeta_6)_{i_2 a_5}^{j_3 i_4}$$
$$+ \tfrac{1}{2} t_{i_1 i_2}^{i_5 i_6} (\zeta_7)_{i_5 i_6}^{j_3 i_4} + \tfrac{1}{2} t_{i_1 i_2}^{a_5 d'_6} (\zeta_8)_{a'_5 d'_6}^{j_3 i_4}$$

single intermediate that should be precomputed is crucial to arriving at an efficient computational sequence. Our work revealed that the costs of evaluating the geminal amplitude equations were $O(n^6)$ for CCSD-R12 and $O(n^7)$ for higher-rank CC-R12, respectively, with n being the number of orbitals. Because the t-amplitude equations of kth-order CC-R12 involve $O(n^{2k+2})$ arithmetic operations, the additional cost incurred by the introduction of the R12 ansatz becomes marginally small for high-rank CC-R12. This suggests that the unapproximated equations of CC-R12 be solved in high-rank CC-R12 for benchmark accuracy.

The algebraic equations and efficient computational sequences were derived by SMITH and reported by us [33] for CCSD-, CCSDT-, and CCSDTQ-R12, their excited-state analogues via the equation-of-motion (EOM) formalisms (EOM-CC-R12 up to EOM-CCSDTQ-R12), and the so-called Λ equations for the analytical gradients and response properties, again up to Λ-CCSDTQ-R12. The full CCSD-, CCSDT-, and CCSDTQ-R12 methods [34,35] were implemented by SMITH into efficient computer codes that took advantage of spin, spatial, and index-permutation symmetries.

The initial benchmark results obtained with the full CCSD-R12 method [34] testified that the various simplified CCSD-R12 methods reported earlier were highly accurate approximations to the full CCSD-R12 method unless the basis set was too small. The assumptions about the relative importance of diagrammatic terms made in these simplified methods were proven to be valid. However, these neglected terms do not increase the computational cost scaling of CCSD-R12 and there appears no need to eliminate them from full CCSD-R12, once they are implemented. In other words, it is important to distinguish whether a certain approximation is motivated by a compromise between accuracy and the computational cost or by that between accuracy and the development cost. The latter has become increasingly unjustifiable with the advent of computerized derivation and implementation.

Figures 2 and 3 plot the valence correlation energies of Ne and FH obtained by various combinations of the CC or CC-R12 methods (using Ten-no's Slater-type correlation function) and basis sets (see ref. 35 for details). These figures are the stunning illustration of the extremely rapid convergence of correlation energies

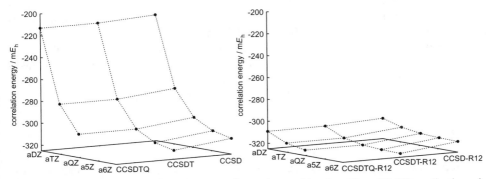

Figure 2 The valence correlation energies of Ne obtained by the CC or CC-R12 methods and the aug-cc-pVXZ basis sets (abbreviated as aXZ) [35].

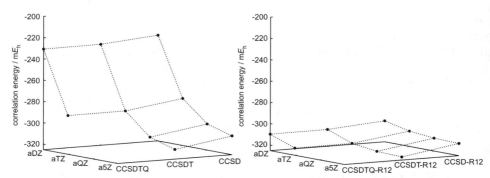

Figure 3 The valence correlation energies of FH obtained by the CC or CC-R12 methods and the aug-cc-pVXZ basis sets (abbreviated as aXZ) [35].

obtained with CC-R12 with respect to both excitation rank and basis-set size. While the basis-set errors observed in CC-R12 decay slightly less rapidly than those in MP2-R12, they are comparable and are expected to display the same $(L+1)^{-7}$ dependence on the basis set that is complete in functions with angular quantum numbers less than or equal to L. This is in stark contrast with the $(L+1)^{-3}$ dependence of the errors in the non-R12 MP2 and CC methods.

The convergence of correlation energies of CC-R12 are sufficiently swift that the complete-correlation, complete-basis-set limits of these polyatomic molecules can be obtained directly, namely, without empirical complete-basis-set extrapolations; with the CC-R12 methods up to CCSDTQ-R12, exact numerical solutions of the Schrödinger equations of polyatomic molecules, an ultimate dream of quantum chemists, appear to be within our reach. In fact, with a grid-based, numerical HF equation solver [78] that could provide the HF energies with 7 or 8 significant figures and CCSD-, CCSDT-, and CCSDTQ-R12 with core electron correlated, we obtained exact (nonrelativistic, Born–Oppenheimer) total energies of Ne, BH, FH, and H_2O at their equilibrium geometries that were -128.9377 ± 0.0004, -25.2892 ± 0.0002, -100.459 ± 0.001, and -76.437 ± 0.003 E_h [35]. With the exception of the H_2O result, the conservative estimates of error bounds were smaller than the chemical accuracy of 1 kcal/mol. These energies also agreed within the quote error bounds with the known experimental values or other accurate calculations that were based on either statistical or empirical treatments. Although we are rarely interested in total energies, our ability to compute them with such high accuracy for general polyatomic molecules has a far-reaching implication on the increasingly greater role of computing, in particular, *ab initio* electron-correlation theory, in chemistry.

5. FUTURE OUTLOOK

A remarkable progress in the CC-R12 methods has been made in the last few years. A variety of approximate, but accurate CCSD-R12 and CCSD(T)-R12 methods as well as full CC-R12 methods through and up to CCSDTQ-R12 have been developed. They constitute perhaps the most rapidly converging series of approximations toward the exact solutions of the Schrödinger equations of a wide variety of polyatomic molecules, for which the HF description is a reasonable zeroth-order approximation. Our group has indeed demonstrated that the total energies of small atoms and molecules could be obtained without empirical extrapolations within a few kilocalories per mole. Apart from an increase in the computational cost by the size-independent factor of 2–10, the CC-R12 methods are vastly more accurate than the non-R12 counterparts and have no significant drawback. We, therefore, expect that the CC-R12 methods will completely replace the CC methods in the near future in all conceivable chemical applications. Among the future research themes in this field are relativistic CC-R12, multi-reference CC-R12, the inclusion of explicit three-electron correlation or CC-R123, analytical energy derivatives and response properties, the ionization and electron-attachment energies, and the extension to large molecules and solids. These research efforts are underway in our laboratories and elsewhere.

ACKNOWLEDGMENTS

We thank Professor Gregory S. Tschumper for the invitation to contribute this review and a critical reading of this manuscript. T. Shiozaki thanks the Japan Society for the Promotion of Science Research Fellowship for Young Scientist and Professor Kimihiko Hirao for his continuous encouragement. E.F. Valeev thanks the Donors of the American Chemical Society Petroleum Research Fund (Grant No. 46811-G6). E.F. Valeev is a Sloan Research Fellow. S. Hirata thanks US Department of Energy (Grant No. DE-FG02-04ER15621) and the Donors of the American Chemical Society Petroleum Research Fund (Grant No. 48440-AC6).

REFERENCES

1. Hirata, S., Yagi, K. Predictive electronic and vibrational many-body methods for molecules and macromolecules. Chem. Phys. Lett. 2008, 464, 123–34.
2. Bartlett, R.J., Musial, M. Coupled-cluster theory in quantum chemistry. Rev. Mod. Phys. 2007, 79, 291–352.
3. Bartlett, R.J. In Modern Electronic Structure Theory II (ed. D.R. Yarkony), World Scientific, Singapore, 1995, pp. 1047–131.
4. Hirata, S., Podeszwa, R., Tobita, M., Bartlett, R.J. Coupled-cluster singles and doubles for extended systems. J. Chem. Phys. 2004, 120, 2581–92.
5. Dunning, T.H. Jr. Gaussian basis sets for use in correlated molecular calculations. I. The atoms boron through neon and hydrogen. J. Chem. Phys. 1989, 90, 1007–23.
6. Almlöf, J., Taylor, P.R. Atomic natural orbital (ANO) basis-sets for quantum-chemical calculations. Adv. Quantum Chem. 1991, 22, 301–73.
7. Pierloot, K., Dumez, B., Widmark, P.-O., Roos, B.O. Density-matrix averaged atomic natural orbital (ANO) basis-sets for correlated molecular wave-functions. IV. Medium-size basis-sets for the atoms H-Kr. Theor. Chim. Acta 1995, 90, 87–114.
8. Feller, D. Application of systematic sequences of wave functions to the water dimer. J. Chem. Phys. 1992, 96, 6104–14.
9. Helgaker, T., Klopper, W., Koch, H., Noga, J. Basis-set convergence of correlated calculations on water. J. Chem. Phys. 1997, 106, 9639–46.
10. Schwenke, D. The extrapolation of one-electron basis sets in electronic structure calculations: how it should work and how it can be made to work. J. Chem. Phys. 2005, 122, 014107.

11. Schwartz, C. Importance of angular correlations between atomic electrons. Phys. Rev. 1962, 126, 1015–19.
12. Kutzelnigg, W., Morgan, J.D. III. Rates of convergence of the partial-wave expansions of atomic correlation energies. J. Chem. Phys. 1992, 96, 4484–508.
13. Kato, T. On the eigenfunctions of many-particle systems in quantum mechanics. Commun. Pure Appl. Math. 1957, 10, 151–77.
14. Pack, R.T., Beyers Brown, W. Cusp conditions for molecular wavefunctions. J. Chem. Phys. 1966, 45, 556–9.
15. Tew, D.P. Second order coalescence conditions of molecular wave functions. J. Chem. Phys. 2008, 129, 014104.
16. Hylleraas, E.A. Neue Berechnung der Energie des Heliums im Grundzustande, sowie des tiefsten Terms von Ortho-Helium. Z. Phys. 1929, 54, 347–66.
17. Hylleraas, E.A. The Schrödinger two-electron atomic problem. Adv. Quantum Chem. 1964, 1, 1–33.
18. Rychlewski, J. (ed.), Explicitly Correlated Wave Functions in Chemistry and Physics, Kluwer Academic Publishers, Dordrecht, 2003.
19. Kutzelnigg, W. r_{12}-Dependent terms in the wave function as closed sums of partial wave amplitudes for large l. Theor. Chim. Acta 1985, 68, 445–69.
20. Valeev, E.F. In Annual Reports in Computational Chemistry (ed. D.C. Spellmeyer), Vol. 2, Elsevier, Amsterdam, 2006, pp. 19–33.
21. Klopper, W., Manby, F.R., Ten-no, S., Valeev, E.F. R12 methods in explicitly correlated molecular electronic structure theory. Int. Rev. Phys. Chem. 2006, 25, 427–68.
22. Helgaker, T., Klopper, W., Tew, D.P. Quantitative quantum chemistry. Mol. Phys. 2008, 106, 2107–43.
23. Kutzelnigg, W., Klopper, W. Wave functions with terms linear in the interelectronic coordinates to take care of the correlation cusp. I. General theory. J. Chem. Phys. 1991, 94, 1985–2001.
24. Klopper, W., Lüthi, H.P. Towards the accurate computation of properties of transition metal compounds: the binding energy of ferrocene. Chem. Phys. Lett. 1996, 262, 546–52.
25. Sinnokrot, M.O., Valeev, E.F., Sherrill, C.D. Estimates of the *ab initio* limit for π-π interactions: the benzene dimer. J. Am. Chem. Soc. 2002, 124, 10887–93.
26. Klopper, W., Samson, C.C.M. Explicitly correlated second-order Moller-Plesset methods with auxiliary basis sets. J. Chem. Phys. 2002, 116, 6397–410.
27. Valeev, E.F. Improving on the resolution of the identity in linear R12 ab initio theories. Chem. Phys. Lett. 2004, 395, 190–5.
28. Ten-no, S. Explicitly correlated second order perturbation theory: introduction of a rational generator and numerical quadratures. J. Chem. Phys. 2004, 121, 117–29.
29. Ten-no, S. Initiation of explicitly correlated Slater-type geminal theory. Chem. Phys. Lett. 2004, 398, 56–61.
30. Kedžuch, S., Milko, M., Noga, J. Alternative formulation of the matrix elements in MP2-R12 theory. Int. J. Quantum Chem. 2005, 105, 929–36.
31. Noga, J., Kutzelnigg, W., Klopper, W. CC-R12, a correlation cusp corrected coupled-cluster method with a pilot application to the Be_2 potential curve. Chem. Phys. Lett. 1992, 199, 497–504.
32. Noga, J., Kutzelnigg, W. Coupled cluster theory that takes care of the correlation cusp by inclusion of linear terms in the interelectronic coordinates. J. Chem. Phys. 1994, 101, 7738–62.
33. Shiozaki, T., Kamiya, M., Hirata, S., Valeev, E.F. Equations of explicitly-correlated coupled-cluster methods. Phys. Chem. Chem. Phys. 2008, 10, 3358–70.
34. Shiozaki, T., Kamiya, M., Hirata, S., Valeev, E.F. Explicitly correlated coupled-cluster singles and doubles method based on complete diagrammatic equations. J. Chem. Phys. 2008, 129, 071101.
35. Shiozaki, T., Kamiya, M., Hirata, S., Valeev, E.F. Higher-order explicitly correlated coupled-cluster methods. J. Chem. Phys. 2009, 130, 054101.
36. Hirata, S. Fast electron-correlation methods for molecular crystals: an application to the α, β_1, and β_2 polymorphs of solid formic acid. J. Chem. Phys. 2008, 129, 204104.
37. Klopper, W., Kutzelnigg, W. Møller–Plesset calculations taking care of the correlation cusp. Chem. Phys. Lett. 1987, 134, 17–22.

38. Dahle, P., Helgaker, T., Jonsson, D., Taylor, P.R. Accurate quantum-chemical calculations using Gaussian-type geminal and Gaussian-type orbital basis sets: applications to atoms and diatomics. Phys. Chem. Chem. Phys. 2007, 9, 3112–26.

39. Neiss, C., Hättig, C. Frequency-dependent nonlinear optical properties with explicitly correlated coupled-cluster response theory using the CCSD(R12) model. J. Chem. Phys. 2007, 126, 154101.

40. Hylleraas, E.A. Über den Grundterm der Zweielektronenprobleme von H^-, He, Li^+, Be^{++} usw. Z. Phys. 1930, 65, 209–25.

41. Tew, D.P., Klopper, W. New correlation factors for explicitly correlated electronic wave functions. J. Chem. Phys. 2005, 123, 074101.

42. Valeev, E.F. Combining explicitly correlated R12 and Gaussian geminal electronic structure theories. J. Chem. Phys. 2006, 125, 244106.

43. Ten-no, S. New implementation of second-order Møller–Plesset perturbation theory with an analytic Slater-type geminal. J. Chem. Phys. 2007, 126, 014108.

44. Janssen, C.L., Nielsen, I.B., Leininger, M.L., Valeev, E.F., Seidl, E.T. MPQC, the massively parallel quantum chemistry program, version 3.0.0α, Sandia National Laboratories, Livermore, CA, 2006.

45. Ahlrichs, R., Bär, M., Baron, H.-P., Bauernschmitt, R., Böcker, S., Crawford, N., Deglmann, P., Ehrig, M., Eichkorn, K., Elliott, S., Furche, F., Haase, F., Häser, M., Hättig, C., Hellweg, A., Horn, H., Huber, C., Huniar, U., Kattannek, M., Köhn, A., Kölmel, C., Kollwitz, M., May, K., Nava, P., Ochsenfeld, C., Öhm, H., Patzelt, H., Rappoport, D., Rubner, O., Schäfer, A., Schneider, U., Sierka, M., Treutler, O., Unterreiner, B., von Arnim, M., Weigend, F., Weis, P., Weiss, H. TURBOMOLE, version 5.9. Universität Karlsruhe, Karlsruhe, Germany, 2006.

46. Werner, H.-J., Knowles, P.J., Lindh, R., Manby, F.R., Schütz, M., Celani, P., Korona, T., Mitrushenkov, A., Rauhut, G., Adler, T.B., Amos, R.D., Bernhardsson, A., Berning, A., Cooper, D.L., Deegan, M. J. O., Dobbyn, A.J., Eckert, F., Goll, E., Hampel, C., Hetzer, G., Hrenar, T., Knizia, G., Köppl, C., Liu, Y., Lloyd, A.W., Mata, R.A., May, A.J., McNicholas, S.J., Meyer, W., Mura, M.E., Nicklass, A., Palmieri, P., Pflüger, K., Pitzer, R., Reiher, M., Schumann, U., Stoll, H., Stone, A.J., Tarroni, R., Thorsteinsson, T., Wang, M., Wolf, A. Molpro, version 2008.1, a package of ab initio programs, Cardiff University, Cardiff, UK, see http://www.molpro.net, 2008.

47. DALTON, A molecular electronic structure program, release 2.0 2005, see http://www.kjemi.uio.no/software/dalton/dalton.html.

48. GELLAN, A hierarchical quantum chemistry program, Nagoya University.

49. Werner, H.-J., Adler, T.B., Manby, F.R. General orbital invariant MP2-F12 theory. J. Chem. Phys. 2007, 126, 164102.

50. Yamaki, D., Koch, H., Ten-no, S. Basis set limits of the second order Møller–Plesset correlation energies of water, methane, acetylene, ethylene, and benzene. J. Chem. Phys. 2007, 127, 144104.

51. Manby, F.R. Density fitting in second-order linear-r_{12} Møller–Plesset perturbation theory. J. Chem. Phys. 2003, 119, 4607–13.

52. Ten-no, S., Manby, F.R. Density fitting for the decomposition of three-electron integrals in explicitly correlated electronic structure theory. J. Chem. Phys. 2003, 119, 5358–63.

53. Werner, H.-J., Manby, F.R. Explicitly correlated second-order perturbation theory using density fitting and local approximations. J. Chem. Phys. 2006, 124, 054114.

54. Werner, H.-J. Eliminating the domain error in local explicitly correlated second-order Møller–Plesset perturbation theory. J. Chem. Phys. 2008, 129, 101103.

55. Valeev, E.F., Janssen, C.L. Second-order Møller–Plesset theory with linear R12 terms (MP2-R12) revisited: auxiliary basis set method and massively parallel implementation. J. Chem. Phys. 2004, 121, 1214–27.

56. Noga, J., Šimunek, J. On the one-particle basis set relaxation in R12 based theories. Chem. Phys. 2009, 356, 1–6.

57. Noga, J., Klopper, W., Kutzelnigg, W. In Recent Advances in Computational Chemistry (ed. R.J. Bartlett), Vol. 3, World Scientific, Singapore, 1997, pp. 1–48.

58. Noga, J., Valiron, P. Explicitly correlated R12 coupled cluster calculations for open shell systems. Chem. Phys. Lett. 2000, 324, 166–74.

59. Noga, J., Valiron, P., Klopper, W. The accuracy of atomization energies from explicitly correlated coupled-cluster calculations. J. Chem. Phys. 2001, 115, 2022–32.

60. Noga, J., Kedžuch, S., Šimunek, J., Ten-no, S. Explicitly correlated coupled cluster F12 theory with single and double excitations. J. Chem. Phys. 2008, 128, 174103.
61. Fliegl, H., Klopper, W., Hättig, C. Coupled-cluster theory with simplified linear-r_{12} corrections: the CCSD(R12) model. J. Chem. Phys. 2005, 122, 084107.
62. Fliegl, H., Hättig, C., Klopper, W. Inclusion of the (T) triples correction into the linear-r_{12} corrected coupled-cluster model CCSD(R12). Int. J. Quantum Chem. 2006, 106, 2306–17.
63. Tew, D.P., Klopper, W., Neiss, C., Hättig, C. Quintuple-ζ quality coupled-cluster correlation energies with triple-ζ basis sets. Phys. Chem. Chem. Phys. 2007, 9, 1921–30.
64. Lee, Y.S., Kucharski, S.A., Bartlett, R.J. A coupled-cluster approach with triple excitations. J. Chem. Phys. 1984, 81, 5906–12.
65. Valeev, E.F. Coupled-cluster methods with perturbative inclusion of explicitly correlated terms: a preliminary investigation. Phys. Chem. Chem. Phys. 2008, 10, 106–13.
66. Torheyden, M., Valeev, E.F. Variational formulation of perturbative explicitly-correlated coupled-cluster methods. Phys. Chem. Chem. Phys. 2008, 10, 3410–20.
67. Valeev, E.F., Crawford, T.D. Simple coupled-cluster singles and doubles method with perturbative inclusion of triples and explicitly correlated geminals: the CCSD(T)$_{R12}$ model. J. Chem. Phys. 2008, 128, 244113.
68. Adler, T.B., Knizia, G., Werner, H.-J. A simple and efficient CCSD(T)-F12 approximation. J. Chem. Phys. 2007, 127, 221106.
69. Tew, D.P., Klopper, W., Hättig, C. A diagonal orbital-invariant explicitly-correlated coupled-cluster method. Chem. Phys. Lett. 2008, 452, 326–32.
70. Bokhan, D., Ten-no, S., Noga, J. Implementation of the CCSD(T)-F12 method using cusp conditions. Phys. Chem. Chem. Phys. 2008, 10, 3320–6.
71. Hirata, S. Tensor contraction engine: abstraction and automated parallel implementation of configuration-interaction, coupled-cluster, and many-body perturbation theories. J. Phys. Chem. A 2003, 107, 9887–97.
72. Hirata, S. Higher-order equation-of-motion coupled-cluster methods. J. Chem. Phys. 2004, 121, 51–9.
73. Hirata, S. Symbolic algebra in quantum chemistry. Theor. Chem. Acc. 2006, 116, 2–17.
74. Kállay, M., Surján, P.R. Higher excitations in coupled-cluster theory. J. Chem. Phys. 2001, 115, 2945–54.
75. Köhn, A., Richings, G.W., Tew, D.P. Implementation of the full explicitly correlated coupled-cluster singles and doubles model CCSD-F12 with optimally reduced auxiliary basis dependence. J. Chem. Phys. 2008, 129, 201103.
76. Kállay, M., Gauss, J., Szalay, P.G. Analytic first derivatives for general coupled-cluster and configuration interaction models. J. Chem. Phys. 2003, 119, 2991–3004.
77. Kállay, M., Gauss, J. Analytic second derivatives for general coupled-cluster and configuration-interaction models. J. Chem. Phys. 2004, 120, 6841–8.
78. Shiozaki, T., Hirata, S. Grid-based Hartree–Fock solutions of polyatomic molecules. Phys. Rev. A 2007, 76, 040503(R).

CHAPTER 7

The Density Matrix Renormalization Group in Quantum Chemistry

Garnet Kin-Lic Chan and **Dominika Zgid**

Abstract

The density matrix renormalization group (DMRG) is an electronic structure method that has recently been applied to *ab initio* quantum chemistry. Even at this early stage, it has enabled the solution of many problems that would previously have been intractable with any other method, in particular, multireference problems with very large active spaces. Here we provide an expository introduction to the theory behind the DMRG and give a brief overview of some recent applications and developments in the context of quantum chemistry.

Keywords: strongly correlated electrons; nondynamic correlation; density matrix renormalization group; post Hartree–Fock methods; many-body basis; matrix product states; complete active space self-consistent field; electron correlation

Department of Chemistry and Chemical Biology, Cornell University, Ithaca, New York, USA

Annual Reports in Computational Chemistry, Volume 5
ISSN: 1574-1400, DOI 10.1016/S1574-1400(09)00507-6

1. INTRODUCTION

The density matrix renormalization group (DMRG) is an electronic structure method that has recently been applied to *ab initio* quantum chemistry. The method originated in the condensed matter community with the pioneering work of White [1,2]. Although the earliest quantum chemistry implementations are only a few years old, the DMRG has already been used to solve many problems that would have been intractable with any other method, and especially, multireference problems with very large active spaces. Unlike a traditional complete active space (CAS) method where the active space wave function is obtained in a brute-force full configuration interaction (FCI) expansion, the DMRG utilizes a compact wave function ansatz. This ansatz is very flexible, is well suited to nondynamic correlation, and in the cases of long molecules, provides a near optimal, local description of multireference correlations.

The current article provides an expository introduction to the DMRG method in quantum chemistry. Here we use a "wave function" point of view that is complementary to earlier articles that work with a primarily renormalization group (RG)–based formulation. The first-time reader will benefit from reading such articles alongside the current one. In addition to this article, we mention again that the DMRG has its origins in the condensed matter community and thus many further excellent sources of information can be found in the physics literature, such as the reviews by Schollwöck [3] and Hallberg [4,5].

The structure of our article is as follows. We begin by introducing the underlying DMRG wave function and examining some of its special properties in Sections 1 and 3. In Section 4, we explain the connection between this wave function ansatz and the algorithmic RG formulation of the DMRG originally proposed by White. In Section 5, we summarize briefly recent applications and developments of the DMRG method in quantum chemistry. We finish with some thoughts and conclusions in Section 6.

2. MOTIVATION FOR THE DMRG ANSATZ

The primary challenge in quantum chemistry is to find a good approximation to the electronic wave function of a quantum state. We can express any N-electron wave function in a complete basis of Slater determinants, through the FCI expansion,

$$|\Psi\rangle = \sum_{\{n\}} \Psi^{n_1 n_2 n_3 \dots n_k} |n_1 n_2 n_3 \dots n_k\rangle \tag{1}$$

$$\{n\} = \{|\text{vac}\rangle, |\uparrow\rangle, |\downarrow\rangle, |\uparrow\downarrow\rangle\} \tag{2}$$

$$\sum_i \{n\} = N \tag{3}$$

Here $|n_1 \ldots n_k\rangle$ is the occupation number representation of the Slater determinant where n_i is the occupation of orbital i. The total number of orbitals is k and N is the total number of electrons.

The dimension of the coefficient tensor Ψ in the above expansion is 4^k, which is intractable for values of k much larger than 10. Therefore, we would like to find an ansatz where Ψ is expressed more compactly. In particular, we would want such an ansatz to require only a *polynomial* amount of information as a function of the number of orbitals in the system, k.

A very simple ansatz would be to approximate the high-dimensional coefficient tensor Ψ by a tensor product of vectors. In elemental form, this would be

$$\Psi^{n_1 n_2 n_3 \ldots n_k} \approx \psi^{n_1} \psi^{n_2} \psi^{n_3} \ldots \psi^{n_k} \tag{4}$$

Note that the vector ψ^n is *not* an orbital vector, but rather a vector of length 4, and ψ^{n_1}, ψ^{n_2} are taken to represent elements of *different* vectors. We can interpret ψ^{n_i} as a variational object associated with the occupancy of orbital i, for example, ψ^{n_1} is associated with the occupancy of the first orbital, ψ^{n_2} with the occupancy of the second, etc. The above product ansatz contains only $4k$ parameters and is certainly very tractable. However, it is also not, in general, very accurate! So, let us try to improve the ansatz by increasing the flexibility of the vectors ψ^n. We can introduce additional *auxiliary* indices, making each vector into a tensor, that is

$$\psi^n \rightarrow \psi^n_{ii'} \tag{5}$$

The new indices i, i' are auxiliary in the sense that they do not appear in the final coefficient tensor Ψ and must be contracted over in some fashion. The simplest arrangement is to contract the indices sequentially from one ψ^n tensor to the next. We then have

$$\Psi^{n_1 n_2 n_3 \ldots n_k} \approx \sum_{i_1 i_2 i_3 \ldots i_{k-1}} \psi^{n_1}_{i_1} \psi^{n_2}_{i_1 i_2} \psi^{n_3}_{i_2 i_3} \ldots \psi^{n_k}_{i_{k-1}} \tag{6}$$

More compactly, we can use matrix notation,

$$\Psi^{n_1 n_2 n_3 \ldots n_k} \approx \psi^{n_1} \psi^{n_2} \psi^{n_3} \ldots \psi^{n_k} \tag{7}$$

where we understand, for example, $\psi^{n_2} \psi^{n_3}$ to denote the matrix product between the two tensors involving the auxiliary indices. For simplicity, we will assume that the dimensions of all auxiliary indices are the same, and we shall call this dimension M. Then the tensors ψ^n are of dimension $4 \times M \times M$ (except for the first and the last) and the total number of parameters in the wave function ansatz is $O(4M^2k)$.

This approximation is, in essence, the DMRG ansatz for M states! (More precisely, it is the ansatz used in the one-site DMRG algorithm). Note that by increasing the dimension M, we can make the ansatz arbitrarily exact. Because the wave function coefficients are obtained as a series of matrix products, the ansatz is also referred to in the literature as the matrix product state [6–9]. Combining the above ansatz for the coefficient tensor explicitly with the Slater determinants yields the full DMRG wave function,

$$|\Psi_{\text{DMRG}}\rangle = \sum_{\{n\}} \psi^{n_1} \psi^{n_2} \psi^{n_3} \ldots \psi^{n_k} |n_1 n_2 n_3 \ldots n_k\rangle \tag{8}$$

3. PROPERTIES OF THE DMRG ANSATZ

Let us now examine some properties of the DMRG ansatz.

1. *Variational*: Since we have an explicit wave function, the expectation value of the energy provides a variational upper bound to the true energy and in practice DMRG energies are evaluated in this way. As M is increased, the DMRG energy converges from above to the exact energy.
2. *Multireference*: There is no division into occupied and virtual orbitals, all orbitals appear on an equal footing in the ansatz (Equation 8). In particular, the Hartree–Fock reference has no special significance here. For this reason, we expect (and observe) the ansatz to be very well balanced for describing nondynamic correlation in multireference problems (see e.g., refs. 10–12). Conversely, the ansatz is inefficient for describing *dynamic* correlation, since to treat dynamic correlation one would benefit from the knowledge of which orbitals are in the occupied and virtual spaces.
3. *Size consistency*: The DMRG ansatz is size-consistent when using a localized basis (e.g., orthogonalized atomic orbitals) in which the wave function for the separated atoms can be considered to factorize into the wave functions for the individual atoms expressed in disjoint subsets of the localized basis. To see this in an informal way, let us assume that we have two DMRG wave functions $|\Psi_A\rangle$ and $|\Psi_B\rangle$ for subsystems A and B separately. Both Ψ_A and Ψ_B have a matrix product structure, that is

$$|\Psi_A\rangle = \sum_{\{n_a\}} \psi^{n_{a1}} \ldots \psi^{n_{ak}} |n_{a1} \ldots n_{ak}\rangle \qquad (9)$$

$$|\Psi_B\rangle = \sum_{\{n_b\}} \psi^{n_{b1}} \ldots \psi^{n_{bk}} |n_{b1} \ldots n_{bk}\rangle \qquad (10)$$

Their product is also a DMRG wave function with a matrix product structure. This then describes the combined system AB in a size-consistent way, that is

$$|\Psi_{AB}\rangle = |\Psi_A\rangle|\Psi_B\rangle = \sum_{\{n_a\}\{n_b\}} \psi^{n_{a1}} \ldots \psi^{n_{ak}} \psi^{n_{b1}} \ldots \psi^{n_{bk}} |n_{a1} \ldots n_{ak}n_{b1} \ldots n_{bk}\rangle \qquad (11)$$

4. *Compactness of the ansatz*: The number of variational parameters in the DMRG ansatz is $O(M^2k)$. How large do we need M to be to achieve a good accuracy? Increasing M increases the correlations that are included in the ansatz. For example, at the $M = 1$ level, the DMRG wave function reduces to a simple product of orbitals in the basis we have chosen. At the $M = 2$ level, the DMRG ansatz can already exactly capture all the correlations in a perfect-pairing valence bond ansatz, and includes some additional correlations between the pairs as well. In general, the M required for a given accuracy depends on the correlations in the specific state of the molecule. However, we have seen for many problems, modest $M = O(100–1,000)$ can allow us to obtain good accuracy. We note that the DMRG incorporates correlations between orbital spaces in a sequential manner, that is, the first set of auxiliary indices i_1 correlates spaces $\{n_1\}$ and $\{n_2 \ldots n_k\}$, i_2 correlates spaces $\{n_1 n_2\}$ and $\{n_3 \ldots n_k\}$, and

so on. For this reason, the DMRG ansatz performs best if strongly correlated orbitals are placed next to each other in the ansatz [13–16]. For example, to recover the perfect-pairing valence bond state in the DMRG $M = 2$ wave function, it is necessary to place the paired orbitals adjacent to each other in the ansatz.

5. *A local multireference ansatz for long molecules*: Because of the sequential nature in which the correlations are incorporated, the DMRG wave function is very well suited to long molecules. In these situations, it can be seen as a naturally local multireference ansatz. In long molecules (i.e., those where one of the dimensions is much larger than the other two), the M required for a given accuracy is *independent* of the length of the system, and the number of variational parameters in the DMRG wave function is simply $C \times O(k)$ where C is a constant, as should be in a local ansatz. However, unlike in other local correlation approaches the DMRG is a local *multireference* ansatz. It is this local nature even in the presence of strong nondynamic correlations, which has allowed us to solve very large active space multireference correlation problems in long molecules [12,17,18].

4. THE CANONICAL FORM AND THE DMRG ALGORITHM

As we have written it in Equation (8), the DMRG wave function contains redundant variational parameters. This means that the set of variational tensors $\psi^{n_1} \ldots \psi^{n_k}$ in the DMRG wave function is not unique, because we can find another set of tensors whose matrix product yields an identical state. This redundancy is analogous to the redundancy of the orbital parametrization of the Hartree–Fock determinant. In the case of the DMRG wave function, we can insert a matrix $\mathbf{T}$ and its inverse between any two variational tensors and leave the state invariant

$$|\Psi\rangle = \sum_{\{n\}} \psi^{n_1} \ldots \psi^{n_k} |n_1 \ldots n_k\rangle$$

$$= \sum_{\{n\}} \psi^{n_1} \ldots \psi^{n_p} \mathbf{T}\mathbf{T}^{-1} \psi^{n_{p+1}} \ldots \psi^{n_k} |n_1 \ldots n_k\rangle$$

$$= \sum_{\{n\}} \psi^{n_1} \ldots \bar{\psi}^{n_p} \bar{\psi}^{n_{p+1}} \ldots \psi^{n_k} |n_1 \ldots n_k\rangle$$

where $\bar{\psi}^{n_p} = \psi^{n_p}\mathbf{T}$, $\bar{\psi}^{n_{p+1}} = \mathbf{T}^{-1}\psi^{n_{p+1}}$, respectively.

To determine an optimal ground state DMRG wave function, we minimize the variational energy $(\Psi|H|\Psi)/(\Psi|\Psi)$. However, because of the redundancy of the variational parameters in the ansatz, it is convenient to work with a *canonical form* of the wave function, where the tensors are uniquely specified by additional constraints. Such a canonical form is intimately associated with the original algorithmic formulation of the DMRG introduced by White [1,2]. The canonical DMRG wave function is written as

$$|\Psi\rangle = \sum_{\{n\}} \mathbf{L}^{n_1} \ldots \mathbf{L}^{n_{p-1}} \mathbf{C}^{n_p} \mathbf{R}^{n_{p+1}} \ldots \mathbf{R}^{n_k} |n_1 \ldots n_k\rangle \qquad (12)$$

Here the $\mathbf{L}^n$ and $\mathbf{R}^n$ renormalization tensors are uniquely specified by the additional orthonormality constraints

$$\sum_n \mathbf{L}^{n\dagger}\mathbf{L}^n = \mathbf{1} \tag{13}$$

$$\sum_n \mathbf{R}^n\mathbf{R}^{n\dagger} = \mathbf{1} \tag{14}$$

We now describe the algorithmic formulation of the DMRG. From the wave function perspective, the DMRG algorithm can be seen simply as an efficient way to optimize the parameters in the canonical DMRG form in Equation (12). However, the algorithm also has an intuitive physical interpretation that is a natural link to the numerical RG of Wilson [19,20] from which it is descended. This is what we now discuss. In this language, the algorithm constructs an optimized, reduced, many-body basis with a special structure. As we see from the explicit ansatz (8) this special structure is of course the matrix product state, but in the original language, the structure arises from the recursive application of *renormalization transformations* to the Hilbert space.

We begin with a formal separation of the Hilbert space of the problem into left and right spaces. Later on, we will see that the renormalization transformations associated with the left spaces are associated with the matrices $\mathbf{L}^n$ in Equation (12) and the renormalization transformations associated with the right spaces are associated with the matrices $\mathbf{R}^n$ in Equation (12). We partition the orbitals $1\ldots k$ into two sets, called the left and right *blocks*. Say the partition is placed at the pth orbital, then orbitals $1\ldots n_{p-1}$ will be part of the left block $\boxed{\text{LEFT}_{1\ldots p-1}}$ and orbitals $n_{p+1}\ldots n_k$ will be part of the right block $\boxed{\text{RIGHT}_{p+1\ldots k}}$. The pth orbital (the partition orbital) is considered separately as part of neither block. We denote this orbital partitioning by the diagram shown in Figure 1.

The orbitals in $\boxed{\text{LEFT}_{1\ldots p-1}}$ span a complete Fock space of dimension 4^{p-1} since every orbital is associated with a Hilbert space of dimension 4 corresponding to the states $|-\rangle, |\uparrow\rangle, |\downarrow\rangle, |\uparrow\downarrow\rangle$ as in Equation (2). Similarly, the orbitals in $\boxed{\text{RIGHT}_{p+1\ldots k}}$ span a complete Fock space of dimension 4^{k-p}. The idea of the DMRG algorithm is to construct a smaller optimized many-body basis $\{l\}$, with a specified reduced dimension M, to span the Fock space of the left block, and a

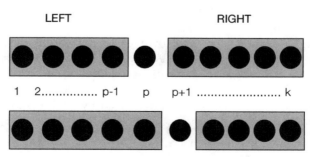

Figure 1 Two steps of the blocking stage of the DMRG algorithm.

similar many-body basis $\{r\}$ of dimension M to span the Fock space of the right block. The small Fock space of the partition orbital p is left untruncated. The product space $\{l\}\otimes\{n_p\}\otimes\{r\}$ (formally of dimension $4M^2$, but suitably restricted to the particle and spin sector of interest) then forms the Hilbert space in which the DMRG wave function is expanded.

We now need to define how the reduced left and right bases, $\{l\}$, $\{r\}$, are obtained. The algorithm to do this operates in a recursive way. When understanding this construction, a useful analogy to keep in mind is proof by induction. If we assume the truth of case p, and prove that this implies case $p+1$, we need to only prove case 1 (i.e., some initial case) for all cases to be true. We shall follow an analogous structure when defining the steps in the DMRG algorithm.

Let us first assume that reduced left and right bases for the left and right blocks, $\boxed{\text{LEFT}_{1\ldots p-1}\,\text{RIGHT}_{p+1\ldots k}}$, (with the partition at orbital p) are available. Projecting the Schrödinger equation onto the product space $\{l\}\otimes\{n_p\}\otimes\{r\}$, we have an effective Schrödinger equation

$$\mathbf{Hc} = E\mathbf{c} \tag{15}$$

where $\mathbf{c}$ are the expansion coefficients of the wave function, that is

$$|\Psi\rangle = \sum_{ln_pr} c_{lnr}|ln_pr\rangle \tag{16}$$

The Hamiltonian matrix in Equation (15) is obtained from appropriate products of representations of second-quantized operators that act within the left block, right block, or partition orbital. For example, in the case of $v_{ijkl}a_i^\dagger a_j^\dagger a_k a_l$, where orbitals ij are in the left partition and orbitals kl are in the right partition, the matrix elements of $a_i^\dagger a_j^\dagger a_k a_l$ are obtained from the representations of $a_i^\dagger a_j^\dagger$ and $a_k a_l$ as

$$\langle ln_pr|a_i^\dagger a_j^\dagger a_k a_l|l'n_p'r'\rangle = \langle l|a_i^\dagger a_j^\dagger|l'\rangle\langle r|a_k a_l|r'\rangle\delta_{n_pn_p'} \tag{17}$$

We assume at this stage that we have all necessary second-quantized operator matrices (associated with $\boxed{\text{LEFT}_{1\ldots p-1}\,\text{RIGHT}_{p+1\ldots k}}$, and the partition orbital) to enable the construction of $\mathbf{H}$ and other physical operators of interest.

Now, let us consider how to proceed to the subsequent orbital partition. This will be called a *renormalization step*. Moving in the left→right direction, the subsequent partition orbital is the $(p+1)$th orbital, and this partitioning corresponds to the second-block configuration in Figure 1. For the recursive reasoning to apply, we must set up our problem in this new partition to resemble that for the previous partition, which means that we need to define how to obtain the reduced M-dimensional left and right bases associated with the new blocks $\boxed{\text{LEFT}_{1\ldots p-1}}$, $\boxed{\text{RIGHT}_{p+1\ldots k}}$, as well as the matrix representations of the necessary second-quantized operators. Temporarily, let us focus only on the expanded left block $\boxed{\text{LEFT}_{1\ldots p}} = \boxed{\text{LEFT}_{1\ldots p-1}} \bullet p$. The Hilbert space of $\boxed{\text{LEFT}_{1\ldots p}}$ is spanned by the product space $\{l\}\otimes\{n_p\}$, (where $\{l\}$ is the reduced basis for the block $\boxed{\text{LEFT}_{1\ldots p-1}}$), but this product space must be reduced in dimension to size M. In other words, we require a projector, or *renormalization transformation* from the product space

$\{l\} \otimes \{n_p\}$ to a *renormalized* basis of M vectors $\{\bar{l}\}$. As argued by White, this projection $\mathbf{L}$ is optimal to target a state vector $\mathbf{c}$, when $\{\bar{l}\}$ are the M largest weight eigenvectors of the statistical density matrix $\mathbf{D}_L$ of the block $\boxed{\text{LEFT}_{1...p}}$ constructed from the target vector $\mathbf{c}$. Writing the coefficient vector as a matrix $\mathbf{C}$ (by assigning the vector element c_{lnr} to the matrix element $C_{(ln)r}$) the statistical density matrix is defined as

$$\mathbf{D}_L = \mathbf{CC}^\dagger \tag{18}$$

and the projector $\mathbf{L}$ is made from the M eigenvectors of the eigenvalue equation

$$\mathbf{D}_L\mathbf{L} = \mathbf{L}(\sigma_1 \ldots \sigma_M)_{\text{diag}}, \quad \sigma_1 \geq \sigma_2 \ldots \geq \sigma_M \tag{19}$$

To complete the definition of the renormalization step for the left block, we also need to construct the new matrix representations of the second-quantized operators. In the product basis $\{l\} \otimes \{n_p\}$, matrix representations can be formed by the product of operator matrices associated with $\boxed{\text{LEFT}_{1...p-1}}$ and the partition orbital p separately. Then, given such a product representation of $\mathbf{O}$ say, the renormalized representation $\bar{\mathbf{O}}$ in the reduced M-dimensional basis $\{\bar{l}\}$ of $\boxed{\text{LEFT}_{1...p}}$ is obtained by projecting with the density matrix eigenvectors $\mathbf{L}$ defined above,

$$\bar{\mathbf{O}} = \mathbf{L}^\dagger \mathbf{OL} \tag{20}$$

We have now defined the procedures in the renormalization step to expand the left block from $\boxed{\text{LEFT}_{1...p-1}}$ to $\boxed{\text{LEFT}_{1...p}}$. We can imagine a sequence of such renormalization steps being applied sequentially left→right, growing the left block from a single orbital until it spans all the orbitals. Such a sequence is known as a renormalization *sweep*. To complete the definition of the sweep (recall proof by induction) we need to define a suitable initialization procedure. We also need to discuss the construction of the right blocks, which we have hitherto assumed known. The initialization procedure is quite simple. Consider the initial left-most partitioning where the left block $\boxed{\text{LEFT}_1}$ corresponds to a single orbital. In this case a suitable basis of dimension $\leq M$ is obtained from the explicit Slater determinants of the left block. The matrix representation of the operators in this basis is easily constructed. To construct the right blocks, we assume that during the initial left→right "warm-up" sweep, only very approximate representations of the right-block reduced basis, consisting of a few randomly chosen low-energy determinants, are used. This does not, of course, constitute an optimal basis representation of the right blocks. However, once the initial warm-up sweep is completed, we carry out a sweep in the reverse direction, that is, growing the right block from $\boxed{\text{RIGHTS}_k}$ to $\boxed{\text{RIGHT}_{1...k}}$. In this reverse sweep, when constructing the renormalized bases for the successive right blocks, the Schrödinger equation is solved in the product basis $\{l\} \otimes \{n_p\} \otimes \{r\}$ where the representations for the left blocks are obtained from the *previous left→right sweep*. In this way, the bases and representations for the right blocks may be improved using a reasonable representation of the left blocks. The left→right and right→left sweeps are then alternated until convergence is achieved. As shown explicitly in ref. 21, the fixed point of the self-consistent optimization occurs when the variational DMRG energy is minimized.

In practice, to carry out all the operations in the DMRG algorithm efficiently, in particular for *ab initio* quantum chemistry Hamiltonians, some care must be put into the optimal organization of intermediate quantities. Such quantum-chemical DMRG algorithms are presented in greater detail in the works discussed in Section 4.

Let us finally see how the DMRG algorithm that we have described above implicitly encodes the canonical DMRG wave function form introduced in Equation (12). The canonical form displayed in Equation (12) refers to the representation of the wave function in the DMRG algorithm when the partition is at the pth orbital. Then the tensor $\mathbf{C}^{n_p}$ is equivalent to the wave function vector $\mathbf{c}$ appearing in the effective Schrödinger Equation (15) through the assignment

$$C_{lr}^n = c_{nlr} \tag{21}$$

The $\mathbf{L}^n$ tensors for the orbitals $1 \ldots p-1$ are obtained from the renormalization transformations $\mathbf{L}$ defining the reduced basis $\{l\}$ of the left block $\boxed{\text{LEFT}_{1 \ldots p-1}}$, by assigning the elements of the eigenvector matrix $\mathbf{L}$ appearing in Equation (19) to the tensor $\mathbf{L}^n$ as $L_{l\bar{l}}^n = L_{(ln)\bar{l}}$. In the same way, the $\mathbf{R}^n$ tensors may be obtained from the renormalization transformations $\mathbf{R}$ defining the reduced basis $\{r\}$. (Note that the orthonormality conditions (13) and (14) are equivalent to the orthonormality conditions satisfied by the eigenvector matrices $\mathbf{L}$ and $\mathbf{R}$.) Explicitly chaining the renormalization transformations $\mathbf{L}$ at each renormalization step together yields the expansion of the renormalized basis state $|l\rangle$ in the determinant basis as

$$|l\rangle = \sum_{\{n\}} (\mathbf{L}^{n_1} \ldots \mathbf{L}^{n_{p-1}})_{1l} |n_1 \ldots n_{p-1}\rangle \tag{22}$$

A similar construction yields the expansion of the right basis states $|r\rangle$

$$|r\rangle = \sum_{\{n\}} (\mathbf{R}^{n_{p+1}} \ldots \mathbf{R}^{n_k})_{r1} |n_{p+1} \ldots n_k\rangle \tag{23}$$

and thus by combining Equations (21)–(23), we observe the equivalence of the canonical matrix product state expression for the DMRG wave function in Equation (12), and the wave function expansion in the renormalized basis of the DMRG algorithm in Equation (16)

$$|\Psi\rangle = \sum_{l\,n\,r} c_{lnr} |ln_p r\rangle \tag{24}$$

$$= \sum_{\{n\}} \mathbf{L}^{n_1} \ldots \mathbf{L}^{n_{p-1}} \mathbf{C}^{n_p} \mathbf{R}^{n_{p+1}} \ldots \mathbf{R}^{n_k} |n_1 \ldots n_k\rangle \tag{25}$$

5. DEVELOPMENTS IN THE DMRG FOR QUANTUM CHEMISTRY

The DMRG method was introduced in 1992 by White [1] and soon was widely applied to problems involving model Hamiltonians in condensed matter. Early applications in conjunction with semi-empirical Hamiltonians focused on the Hubbard and Pariser–Parr–Pople (PPP) models for conjugated systems (see e.g., refs. 22–28). As a representative example, we consider the work by Fano et al. [25] who performed DMRG studies on cyclic polyenes ($C_M H_M$, $M = 4m + 2$, $m = 1, 2, \ldots$) (also known as annulenes) in the PPP model [29–31]. These calculations obtained an excellent, near-exact treatment of the electronic correlation. Given the well-known difficulties of single-reference quantum chemistry methods such as coupled cluster theory to describe correlation in these systems (see e.g., studies by Paldus and coworkers [32,33]), such early work demonstrated the promise of the DMRG method for multireference chemistry problems.

In 1999, White and Martin [34] carried out the first application of the DMRG method to a molecule using the full *ab initio* quantum-chemical Hamiltonian. In particular, they demonstrated that a DMRG calculation with a moderate number of states could recover near FCI accuracy for the water molecule in a double-zeta with polarization basis. The White and Martin paper was significant also in formulating the intermediates that are necessary for an efficient implementation of the DMRG in quantum chemistry. After these initial papers, other groups in quantum chemistry started to investigate the DMRG method. Currently, several implementations exist, including that of Mitrushenkov et al. [35], ours [13], that of Legeza, Röder, and Hess [36], and more recently, Zgid and Nooijen [37], and Kurashige and Yanai [38].

Over the last few years, the DMRG has been applied to a wide variety of molecular problems in quantum chemistry. In the early stages, these were typically small molecular problems for which benchmark molecular results were available. Some examples include the singlet–triplet gap in HHeH [39], benchmark water and nitrogen curves in small basis sets [13,40], singlet–triplet gaps in methylene [37], and the ionic–covalent curve-crossing in LiF [41]. More recently, thanks to efficient implementations, it has become possible to use the DMRG to obtain near-exact solutions to molecular problems considerably beyond the capabilities of FCI. Some examples of this include our exact solution of the Schrödinger equation for water in a triple-zeta, double-polarization basis [42], as well as for the nitrogen-binding curve at the all-electron double-zeta with polarization basis level [11]. However, while the use of the DMRG to obtain near-exact solutions to the Schrödinger equation for problems beyond conventional FCI is of course interesting, the early molecular studies highlighted a more significant point, namely the types of correlation that the DMRG wave function is efficient at capturing. For example, in our studies of the nitrogen-binding curve, we compared DMRG calculations to high-level coupled cluster with up to hextuple excitations (CCS-DTQPH). While the CCS-DTQPH energies and the DMRG energies were comparable at the equilibrium geometry, the DMRG energies for a fixed number of states M retained essentially the same accuracy across the entire potential energy curve. This and other studies demonstrated the ability of the DMRG wave function to capture multireference correlation in a balanced way, as we described in Section 3. Conversely, when moving from a small basis to a larger basis DMRG calculation for the same molecule (e.g., from a double-zeta to a triple-zeta basis for the water molecule, as in ref. 13), the number of states that need to be kept in the DMRG ansatz to achieve a given accuracy had to be increased significantly, demonstrating that dynamical correlation is not efficiently captured by the DMRG wave function. Thus, we view the most promising domain of application of the DMRG method must to be to solve the *active space* multireference problem. With current DMRG technology, a nearly exact treatment of the CAS correlation for arbitrary molecules with up to roughly 30 active orbitals and electrons can be obtained.

Given the strength of the DMRG method for large-scale multireference electronic structure, a clear domain of application must be to complicated transition metal problems. Although such applications are still at a relatively early stage, the Reiher group has performed some preliminary studies [15,43–46]. For example, they used the DMRG method [46] to calculate the spin-gap of the Cu_2O_2 core of tyrosinase, a problem that had evaded conventional CAS methods due to the need for a large active space.

More recently Kurashige and Yanai [38], in a demonstration of their highly efficient DMRG implementation, not only obtained correctly converged DMRG energies for the same Cu_2O_2 system that had been studied earlier but not fully converged by Reiher group, but also carried out an impressive near-exact solution of the CAS problem for the Cr_2 molecule correlating an active space of 24 electrons in 30 orbitals.

One of the directions of our own group in recent years has been to use the DMRG as an efficient local multireference method. This will be the case when the systems we are considering are large in one of their dimensions, that is, long molecules. In these ideal settings, the DMRG method obtains near-exact active space solutions of the Schrödinger equation for problem sizes completely inconceivable using other techniques, for example, for 100-orbital, 100-electron active spaces. In our first demonstration, we showed how the DMRG could exactly describe the simultaneous bond breaking of 49 bonds in a hydrogen chain, a problem nominally requiring a 50-electron, 50-orbital active space. In more recent works, we have used our local DMRG method to study excited states in conjugated systems, which have significant multireference character, ranging from unusual polyradical character in the acene series [18], to magnetism in polyphenylcarbenes [47], to the nature of dark states in light-harvesting pigments [48].

The application of the DMRG to the problems of quantum chemistry has of course necessitated new theoretical developments related to the DMRG itself. A concern in many early studies and applications to molecular problems was the problem of orbital ordering, since, as discussed in Section 4, the sequential incorporation of correlations requires that strongly correlated orbitals are placed near each other in the DMRG ansatz. There have been a number of different proposals for this problem (see refs. 13–16, 35). It is clear that there can be no (efficient) algorithm that yields the optimal orbital ordering in *every* case, but certainly one wishes to find a scheme that obtains a reasonable ordering most of the time. The solution to this problem, however, is to our minds, still open.

There have also been many theoretical developments that have extended the applicability and functionality of the DMRG method for quantum chemistry. Some have been algorithmic nature, for example, efficient parallel algorithms

[10], spin-adapted formulations [37], the evaluation of the two-body density matrix within DMRG method [49], dynamical selection of the number of DMRG states to use in the calculation [36], formulations for non-orthogonal basis sets [50,51], improved algorithms for excited states [17], and reduced-scaling methods for large molecules [18]. The recent developments, which we believe will be particularly useful for quantum chemistry, are the development of analytic derivatives [52], and the implementation of orbital optimization schemes [48,53]. In particular, the latter allows the DMRG to serve as a replacement for current CASSCF methodologies, for systems where multireference correlation must be treated in a large active space.

6. CONCLUSIONS

In this article, we have attempted to introduce the DMRG from the viewpoint of quantum chemistry. Our emphasis has been on the wave function formulation of the DMRG, which is most natural for chemical applications. The DMRG provides a new approach to the problem of multireference correlation due to the use of optimal many-body basis states with a matrix product structure, in contrast to traditional approaches that are expressed explicitly in the basis of Slater determinants. We have highlighted in our presentation some favorable properties of the DMRG ansatz, such as its variational and multireference nature, compactness, and size consistency, as well its natural locality. Additionally, we have discussed the relationship between the abstract DMRG ansatz and its efficient realization in the original DMRG algorithm. Finally, we have given a brief overview of recent applications of the DMRG method to chemical systems, focusing in particular on cases where the method has made possible the treatment of previously inaccessible problems involving very large active spaces.

ACKNOWLEDGMENTS

Garnet Kin-Lic Chan would like to acknowledge support from Cornell University, the Cornell Center for Materials Research (CCMR), the David and Lucile Packard Foundation, the National Science Foundation CAREER Program CHE-0645380, the Alfred P. Sloan Foundation, and the Department of Energy, Office of Science through award DE-FG02-07ER46432.

REFERENCES

1. White, S.R. Density matrix formulation for quantum renormalization groups. Phys. Rev. Lett. 1992, 69(19), 2863.
2. White, S.R. Density-matrix algorithms for quantum renormalization groups. Phys. Rev. B 1993, 48(14), 10345.
3. Schollwöck, U. The density-matrix renormalization group. Rev. Mod. Phys. 2005, 77(1), 259.
4. Hallberg, K. Density matrix renormalization: a review of the method and its applications. In Theoretical Methods for Strongly Correlated Electrons (eds D. Sénéchal, A.-M. Tremblay, and C. Bourbonnais), CRM Series in Mathematical Physics. Springer, New York, 2003.

5. Hallberg, K.A. New trends in density matrix renormalization. Adv. Phys. 2006, 55(5), 477.
6. Fannes, M., Nachtergaele, B., Werner, R.F. Finitely correlated states on quantum spin chains. Comm. Math. Phys. 1992, 144(3), 443.
7. Fannes, M., Nachtergaele, B., Werner, R.F. Finitely correlated pure states. J. Funct. Anal. 1994, 120(2), 511–34.
8. Östlund, S., Rommer, S. Thermodynamic limit of density matrix renormalization. Phys. Rev. Lett 1995, 75(19), 3537.
9. Rommer, S., Östlund, S. Class of ansatz wave functions for one-dimensional spin systems and their relation to the density matrix renormalization group. Phys. Rev. B 1997, 55(4), 2164.
10. Chan, G.K.L. An algorithm for large scale density matrix renormalization group calculations. J. Chem. Phys. 2004, 120(7), 3172.
11. Chan, G.K.L., Kállay, M., Gauss, J. State-of-the-art density matrix renormalization group and coupled cluster theory studies of the nitrogen binding curve. J. Chem. Phys. 2004, 121(13), 6110.
12. Hachmann, J., Cardoen, W., Chan, G.K.L. Multireference correlation in long molecules with the quadratic scaling density matrix renormalization group. J. Chem. Phys. 2006, 125(14), 144101.
13. Chan, G.K.L., Head-Gordon, M. Highly correlated calculations with a polynomial cost algorithm: a study of the density matrix renormalization group. J. Chem. Phys. 2002, 116(11), 4462.
14. Legeza, Ö., Sólyom, J. Optimizing the density-matrix renormalization group method using quantum information entropy. Phys. Rev. B 2003, 68(19), 195116.
15. Moritz, G., Hess, B.A., Reiher, M. Convergence behavior of the density-matrix renormalization group algorithm for optimized orbital orderings. J. Chem. Phys. 2005, 122(2), 024107.
16. Rissler, J., Noack, R.M., White, S.R. Measuring orbital interaction using quantum information theory. Chem. Phys. 2006, 323(2–3), 519.
17. Dorando, J.J., Hachmann, J., Chan, G.K.L. Targeted excited state algorithms. J. Chem. Phys. 2007, 127(8), 084109.
18. Hachmann, J., Dorando, J.J., Avilés, M., Chan, G.K.L. The radical character of the acenes: a density matrix renormalization group study. J. Chem. Phys. 2007, 127(13), 134309.
19. Wilson, K.G. The renormalization group: critical phenomena and the kondo problem. Rev. Mod. Phys. 1975, 47(4), 773.
20. Wilson, K.G. The renormalization group and critical phenomena. Rev. Mod. Phys. 1983, 55(3), 583.
21. Chan, G.K.L. Density matrix renormalisation group Lagrangians. Phys. Chem. Chem. Phys. 2008, 10, 3454.
22. Ramasesha, S., Pati, S.K., Krishnamurthy, H.R., Shuai, Z., Brédas, J.L. Low-lying electronic excitations and nonlinear optic properties of polymers via symmetrized density matrix renormalization group method. Synth. Met. 1997, 85(1–3), 1019.
23. Yaron, D., Moore, E.E., Shuai, Z., Brédas, J.L. Comparison of density matrix renormalization group calculations with electron-hole models of exciton binding in conjugated polymers. J. Chem. Phys. 1998, 108(17), 7451.
24. Shuai, Z., Brédas, J.L., Saxena, A., Bishop, A.R. Linear and nonlinear optical response of polyenes: a density matrix renormalization group study. J. Chem. Phys. 1998, 109(6), 2549.
25. Fano, G., Ortolani, F., Ziosi, L. The density matrix renormalization group method: Application to the PPP model of a cyclic polyene chain. J. Chem. Phys. 1998, 108(22), 9246.
26. Bendazzoli, G.L., Evangelisti, S., Fano, G., Ortolani, F., Ziosi, L. Density matrix renormalization group study of dimerization of the pariserparrpople model of polyacetilene. J. Chem. Phys. 1999, 110(2), 1277.
27. Raghu, C., Anusooya Pati, Y., Ramasesha, S. Structural and electronic instabilities in polyacenes: density-matrix renormalization group study of a long-range interacting model. Phys. Rev. B 2002, 65(15), 155204.
28. Raghu, C., Anusooya Pati, Y., Ramasesha, S. Density-matrix renormalization-group study of low-lying excitations of polyacene within a Pariser–Parr–Pople model. Phys. Rev. B 2002, 66(3), 035116.
29. Pariser, R., Parr, R. A semi-empirical theory of the electronic spectra and electronic structure of complex unsaturated molecules, I. J. Chem. Phys. 1953, 21, 466.
30. Pariser, R., Parr, R. A semi-empirical theory of the electronic spectra and electronic structure of complex unsaturated molecules. II. J. Chem. Phys. 1953, 21, 767.

31. Pople, J.A. Electron interaction in unsaturated hydrocarbons. Trans. Faraday Soc 1953, 49, 1375.
32. Paldus, J., Takahashi, M., Cho, R.W.H. Coupled-cluster approach to electron correlation in one dimension: cyclic polyene model in delocalized basis. Phys. Rev. B 1984, 30, 4267.
33. Paldus, J., Čížek, J., Takahashi, M. Approximate account of the connected quadruply excited clusters in the coupled-pair many-electron theory. Phys. Rev. A 1984, 30, 2193–209.
34. White, S.R., Martin, R.L. Ab initio quantum chemistry using the density matrix renormalization group. J. Chem. Phys. 1999, 110(9), 4127.
35. Mitrushenkov, A.O., Fano, G., Ortolani, F., Linguerri, R., Palmieri, P. Quantum chemistry using the density matrix renormalization group. J. Chem. Phys. 2001, 115(15), 6815.
36. Legeza, Ö., Röder, J., Hess, B.A. Controlling the accuracy of the density-matrix renormalization-group method: the dynamical block state selection approach. Phys. Rev. B 2003, 67(12), 125114.
37. Zgid, D., Nooijen, M. On the spin and symmetry adaptation of the density matrix renormalization group method. J. Chem. Phys. 2008, 128, 014107.
38. Kurashige, Y., Yanai, T. High-performance *ab initio* density matrix renormalization group method: applicability to large-scale multireference problems for metal compounds. J. Chem. Phys. 2009, 130(23), 234114.
39. Daul, S., Ciofini, I., Daul, C., White, S.R. Full-CI quantum chemistry using the density matrix renormalization group. Int. J. Quantum Chem. 2000, 79(6), 331.
40. Mitrushenkov, A.O., Linguerri, R., Palmieri, P., Fano, G. Quantum chemistry using the density matrix renormalization group II. J. Chem. Phys. 2003, 119(8), 4148.
41. Legeza, Ö., Röder, J., Hess, B.A. Qc-DMRG study of the ionic-neutral curve crossing of life. Mol. Phys. 2003, 101(13), 2019.
42. Chan, G.K.L., Head-Gordon, M. Exact solution (within a triple-zeta, double polarization basis set) of the electronic Schrödinger equation for water. J. Chem. Phys. 2003, 118(19), 8551.
43. Moritz, G., Reiher, M. Construction of environment states in quantum-chemical density-matrix renormalization group calculations. J. Chem. Phys. 2006, 124(3), 034103.
44. Moritz, G., Wolf, A., Reiher, M. Relativistic DMRG calculations on the curve crossing of cesium hydride. J. Chem. Phys. 2005, 123(18), 184105.
45. Moritz, G., Reiher, M. Decomposition of density matrix renormalization group states into a Slater determinant basis. J. Chem. Phys. 2007, 126(24), 244109.
46. Marti, K.H., Ondík, I.M., Moritz, G., Reiher, M. Density matrix renormalization group calculations on relative energies of transition metal complexes and clusters. J. Chem. Phys. 2008, 128, 014104.
47. Yanai, T., Kurashige, Y., Ghosh, D., Chan, G.K-L. Accelerating convergence in iterative solutions of large active-space self-consistent field calculations. Int. J. Quantum Chem., 2009, 109, 2178–90.
48. Ghosh, D., Hachmann, J., Yanai, T., Chan, G.K.L. Orbital optimization in the density matrix renormalization group, with applications to polyenes and β-carotene. J. Chem. Phys. 2008, 128, 144117.
49. Zgid, D., Nooijen, M. Obtaining the two-body density matrix in the density matrix renormalization group method. J. Chem. Phys. 2008, 128, 144115.
50. Mitrushenkov, A.O., Fano, G., Linguerri, R., Palmieri, P. On the possibility to use non-orthogonal orbitals for density matrix renormalization group calculations in quantum chemistry. arXiv.cond-mat, 0306058v1, 2003.
51. Chan, G.K.L., Van Voorhis, T. Density-matrix renormalization-group algorithms with nonorthogonal orbitals and non-Hermitian operators, and applications to polyenes. J. Chem. Phys. 2005, 122(20), 204101.
52. Dorando, J.J., Hachmann, J., Chan G.K-L. Analytic response theory for the density matrix renormalization group. J. Chem. Phys. 2009, 130(18), 184111.
53. Zgid, D., Nooijen, M. The density matrix renormalization group self-consistent field method: Orbital optimization with the density matrix renormalization group method in the active space. J. Chem. Phys. 2008, 128, 144116.

CHAPTER 8

Electron Transfer in Gaseous Positively Charged Peptides — Relation to Mass Spectrometry

Jack Simons

Contents

Abstract

Special theoretical tools are needed to carry out *ab initio* simulations of (i) electron transfer from a negatively charged donor (i.e., an anion donor) to a positively charged polypeptide and (ii) electron transfer within such a peptide from Rydberg orbitals on positive sites (e.g., protonated amines on side chains) to disulfide or amide bond sites. Basis sets capable of describing several Rydberg states as well as states with an electron attached to an SS σ^* or OCN π^* orbital must be used. Electron correlation is important to include for some states, and methods that allow one to obtain excited states of the same spin and spatial symmetry must be employed. Tools for treating surface hopping between states are also crucial. Examples of applying such tools to anion-to-peptide and intra-peptide electron-transfer processes are presented. It is demonstrated that intra-peptide electron transfer from Rydberg orbitals can occur over long distances (15 Å) and can take place in

Chemistry Department and Henry Eyring Center for Theoretical Chemistry, University of Utah, Salt Lake City, UT 84112, USA

Annual Reports in Computational Chemistry, Volume 5
ISSN: 1574-1400, DOI 10.1016/S1574-1400(09)00508-8

both through-space and through-bond paths. Similarities and differences with other electron-transfer processes in chemistry are also discussed.

Keywords: electron-capture dissociation; electron-transfer dissociation; electron transfer; Rydberg orbital; Landau–Zener theory

1. INTRODUCTION

Electron-capture dissociation (ECD) [1] and electron-transfer dissociation (ETD) [2] mass spectroscopic methods have shown much utility and promise for sequencing peptides and proteins. A strong point of both techniques is their propensity for selectively cleaving disulfide and N–C_α bonds and for doing so over a wide range of the backbone, thus producing many different fragment ions, unlike collision-induced dissociation (CID) or infrared multiphoton dissociation (IRMPD). ECD and ETD also preserve labile sidechains with posttranslational modifications. Parallel with many advances in the experimental development and improvement of these methods, theoretical studies have been carried out by several groups to try to determine the mechanism(s) [3] by which electron attachment leads to these specific bond cleavages as well as how the initial electron attachment occurs.

In both ECD and ETD experimental approaches, a positively charged sample of a polypeptide enters the gas phase (usually via electrospray), after which ions of specific mass to charge ratio are selected. Usually, the positive charging is induced by subjecting the solution-phase sample to acidic conditions prior to electrospray. An example of a relatively simple polypeptide is shown in Figure 1 as a means for introducing several concepts and terminology.

In ETD, an anion donor collides with the positively charged peptide and transfers an electron to the peptide; subsequent to this intermolecular electron transfer, the peptide undergoes cleavage at one of its N–C_α or S–S bonds to form fragment ions. The mass to charge ratios and intensities of the fragment ions are the raw data that is then used to infer the primary sequence of the original

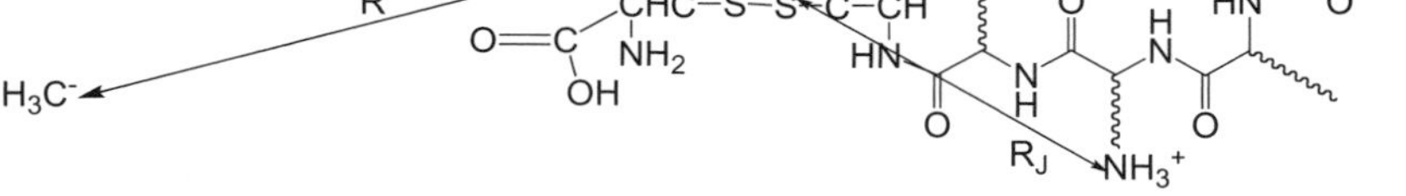

Figure 1 Prototypical polypeptide showing disulfide (SS) linkage, one of many N–C_α bonds, amino acid side chains (wavy lines), protonated amines on side chains (wavy lines), and one of many peptide bonds. Also shown is an anion donor (H_3C^-) colliding with the peptide.

polypeptide. In ECD, a free electron (usually having low kinetic energy) rather than a molecular anion collides with the parent polypeptide. This electron is captured and subsequently the peptide undergoes cleavage at one of its N–C_α or S–S bonds. The kind of fragment ions produced (i.e., those arising from N–C_α or S–S bond cleavage) and their intensities are found to be very similar for ETD and ECD, suggesting that the two processes proceed along very similar mechanistic paths. The detailed mechanism(s) by which the electron attaches to the peptide, where it attaches, and how the N–C_α or S–S bond cleavage then takes place have been the main focuses of our research in this area.

1.1 The electron-capture event involves electron transfer

In both ECD and ETD, the initial conditions appropriate to the experiments do not correspond to the ground electronic state of the electron/peptide (ECD) or anion/peptide (ETD) system. In both cases, there are a myriad of lower-energy electronic states, and this fact presents major challenges to the theoretical study of these processes. In Figure 2, we show qualitative plots of energies as functions of the distance R between a H_3C^- anion donor and a polypeptide having total charge Z.

The families of electronic states that must be considered in such a study and that are depicted in Figure 2 include:

1. The ion-pair state in which the "excess" electron resides on the donor anion; this state's energy varies strongly with R reflecting the strong Coulomb attraction between the anion donor and the positively charged polypeptide. In Figure 2, this state is shown as rapidly descending as R decreases approximately as expected based on the Coulomb attraction between the anion donor and the peptide of charge Z: $-14.4Z/R$ is in eV, when R is in Å.
2. Families of Rydberg states in which the excess electron has moved from the anion donor to reside in a Rydberg orbital (ground 3s, or excited 3p, 3d, 4s, etc.) on one of the polypeptide's protonated amine side chains. These curves (at least

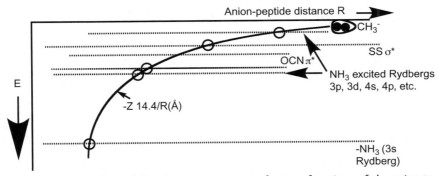

Figure 2 Qualitative plots of the electronic energy surfaces as functions of the anion-to-peptide distance R, for the anion–peptide collision complex, and for states in which the electron has been transferred from the anion to Rydberg states on one of the peptide's protonated amines, to an SS σ^* orbital, or to an amide π^* orbital.

at long anion–peptide distances) are found to vary rather weakly with R because the anion donor has been rendered neutral, so only charge–dipole and charge-induced-dipole potentials between the peptide and the H_3C radical exist.
3. One or more states in which the excess electron has moved to reside in an antibonding SS σ^* orbital of one of the peptide's disulfide linkages.
4. One or more states in which the excess electron has moved to reside in an antibonding OCN π^* orbital of one of the peptide's amide linkages. The curves of these σ^* and π^* vary rather weakly with R for the same reasons as noted above.

Near where we depict the energy surfaces crossing in Figure 2, the pairs of surfaces actually undergo avoided crossings at which they experience a minimum energy splitting that we denote $2H_{1,2}$. Moving through each such avoided crossing, the nature of the two states changes. For example, when the ion-pair state approaches the $-NH_3$ 3s ground-Rydberg state from above at the left-most circle in Figure 2, the lower-energy surface corresponds to having the extra electron in the 3s Rydberg orbital; the upper surface has this electron in the methyl lone pair orbital. In contrast, to the left of the circle, the lower surface corresponds to the ion-pair state, while the upper surface is the 3s Rydberg-attached state. The evolution of the two states' energies and wave functions through such avoided crossings describes how the interspecies electron transfer occurs. This is the *first category of electron-transfer processes* one needs to study when investigating ETD or ECD.

In probing ETD experiments, one must be able to characterize the above four families of electronic energy surfaces, and one must have a means of extracting the couplings $H_{1,2}$ between these states as they undergo avoided crossings. In the studies that our group has undertaken [3h–3w], we have used Landau–Zener (LZ) theory to estimate the probabilities P for an electron being transferred from an anion donor to a Rydberg orbital, an SS σ^* orbital, or an amide π^* orbital during a collision beginning on the attractive ion-pair surface that undergoes a crossing with one of the other surfaces. In LZ theory, this probability is computed as

$$P = 1 - \exp\left[-\frac{2\pi H_{1,2}^2}{hv|\Delta F|}\right] \approx \frac{2\pi H_{1,2}^2}{hv|\Delta F|} \tag{1}$$

$H_{1,2}$ is one half the splitting observed when the two energy surfaces undergo their avoided crossing, v the speed at which the lion pair moves through the avoided crossing region, and ΔF the difference in the slopes of the two energy surfaces as they approach the avoided crossing.

1.2 Intra-peptide electron transfer can also occur

Once an electron is transferred to or captured by the polypeptide, various things can happen:

1. If the electron attaches directly to an SS σ^* orbital, the disulfide bond promptly cleaves [3j] because the $\sigma^2\sigma^{*1}$ electron-attached state is strongly

repulsive along the S–S bond. This is one path by which disulfide cleavage occurs.

2. If the electron enters an OCN π^* orbital, an $^-$O–C•–NH–C$_\alpha$ radical anion center is formed, after which the neighboring N–C$_\alpha$ bond is weakened and can be cleaved (to produce $^-$O–C$=$NH + •C$_\alpha$) thus producing the N–C$_\alpha$ bond-cleavage products [3m].

3. If the electron enters a Rydberg orbital on one of the protonated amine sites, in addition to undergoing a cascade of radiative or non-radiative relaxation steps to lower-energy Rydberg states, it can subsequently undergo intra-peptide electron transfer to either an SS σ^* or an OCN π^* orbital after which disulfide or N–C$_\alpha$ bond cleavage can occur [3r,3u–3w].

For the intra-peptide electron migration to be effective in cleaving an S–S or N–C$_\alpha$ bond, it must occur before the Rydberg species from which the electron is transferred can decay by some other mechanism. It is believed that electron attachment (in ECD or ETD) at a positively charged side chain initially occurs into an excited-Rydberg orbital after which a decay cascade eventually leads to formation of the ground-Rydberg species. It is known that excited-Rydberg states belonging to protonated or fixed-charge amine sites undergo radiationless relaxation to the ground-Rydberg state in a few to several microseconds. Moreover, we know that the excited-Rydberg states do not undergo N–H or N–C bond cleavage, but the ground-Rydberg states do (in *ca.* 10^{-12} s). Hence, the intra-peptide electron transfer must occur within a few microseconds of the time the electron attaches to an excited-Rydberg orbital; otherwise, it will relax to the ground-Rydberg state and N–H or N–C bond cleavage will occur (ejecting an H atom or an alkyl radical) terminating the electron's chance to undergo further transfer.

This transfer from a Rydberg orbital to an SS or OCN antibonding orbital is the *second family of electron-transfer events* that must be considered when studying ECD or ETD. These transfers can occur either through-space or through-bond. To appreciate which Rydberg states are most likely to be involved, qualitative depictions of the energies of states in which the extra electron occupies a Rydberg orbital or an SS σ^* orbital are shown in Figure 3 as functions of the S–S bond length.

The energy profile of the SS σ^*-attached state is largely repulsive,[1] but its location, relative to the parent and Rydberg-attached states, depends upon the distance R between the SS bond and the positively charged site whose Coulomb potential acts to move the SS σ^*-attached state up and down in energy as R varies. For example, if R is very large, the energy of the SS σ^*-attached state will be little affected by the stabilizing Coulomb potential of the $-NH_3^+$ site and thus its energy profile will be as shown by the upper curve in Figure 3. Alternatively, if the $-NH_3^+$ site is closer to the SS bond, the energy profile will be shifted downward as in the lower curve in Figure 3.

For each instantaneous value of the Coulomb potential experienced by the SS σ^* orbital, a different Rydberg state will intersect the energy profile of the

---

[1]This state's energy is weakly attractive at large distances because of van der Waals and charge-induced dipole interactions, but its valence-range character is repulsive.

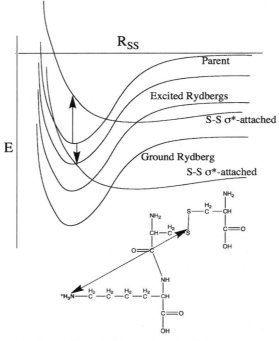

Figure 3 Energies, as functions of the S–S bond length, of the parent charged polypeptide (top), of ground and excited-Rydberg states localized on the protonated amine side chain, and of the SS σ*-attached state in the absence of (upper curve) and in the presence of (lower curve) Coulomb stabilization (appears as Figure 1 in ref. 3s).

SS σ*-attached state at or near the equilibrium SS bond length R_e. In poly-peptides containing multiple positively charged sites such as that shown in Figure 4, the total Coulomb potential C

$$C = -14.4 \sum_J \frac{1}{R_J} \qquad (2)$$

will determine the energy-placement of the SS σ*-attached state (R_J is the distance of the Jth charged site to the SS bond).

Because ETD and ECD experiments are carried out at or near room temperature, the SS and N–C$_\alpha$ bonds are expected to sample only distances close to their equilibrium values R_e. Hence, we focus primarily on the Rydberg states having energies close to that of the SS σ*-attached or OCN π*-attached state near R_e when considering intra-peptide electron transfer. In Figure 3, this would be the highest Rydberg state shown.

In the studies our group has undertaken [3h–3w] to date, we used LZ theory to estimate the probabilities P for an electron being transferred from such a Rydberg orbital to an SS σ* or amide π* orbital. In Figure 5 we show actual data from such a study on the $H_3C-S-S-(CH_2)_3-NH_3^+$ model compound.

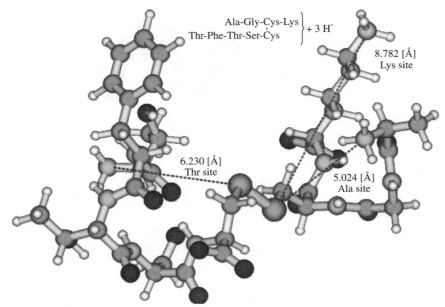

Figure 4 Triply protonated polypeptide containing one SS linkage with the distances R_J to each positive site labeled by dotted lines (appears in Figure 7 of ref. 3s).

From the data shown in Figure 5, we concluded that it is the excited-Rydberg state that crosses the repulsive SS σ^*-attached state near R_e, so this is the state from which electron transfer is most likely to occur. The 82 cm^{-1} energy value shown in Figure 5 is the electronic coupling matrix element $H_{1,2}$ connecting the excited-Rydberg and SS σ^* states, which plays a central role in determining the LZ-estimated probability P of electron transfer (see Equation (1)). In these cases, the rates of electron transfer are computed by multiplying the frequency v at which the S–S bond moves through the curve crossing (we take this to be the harmonic frequency of the SS bond) by the LZ probability P. In the LZ formula, the speed v at which the system passes through the crossing region is computed in terms of the speed of the SS vibrational motion.

To illustrate, it was shown in ref. 3q that $H_{1,2}$ values in the 300 cm^{-1} range produce LZ probabilities of $ca.$ 0.1–0.5 for this system. Thus, we can estimate the rates of electron transfer by multiplying the S–S vibrational frequency v_{SS} ($ca.$ 1.5×10^{13} s^{-1}) by the surface hopping probability (0.1–0.5) and then scaling by the ratio of the square of ($H_{1,2}/300$):

$$\text{Rate} \approx (1.5 \text{ to } 7.5) \times 10^{12} \left[\frac{H_{1,2}}{300}\right]^2 \text{s}^{-1} \tag{3}$$

Such estimates allowed us to conclude that the smallest $H_{1,2}$ that could produce S–S bond cleavage competitive with relaxation from one Rydberg state to another (taking place at $ca.$ 10^6 s^{-1}) should be $H_{1,2}^{min} \approx 0.11 - 0.24$ cm^{-1}. Most of the $H_{1,2}$ values we obtained in our studies to date are substantially larger,

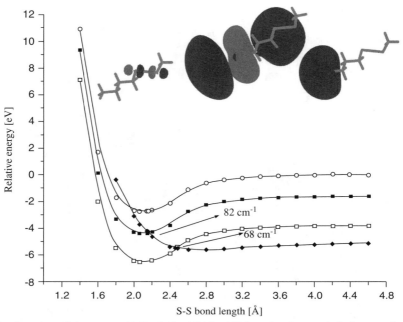

Figure 5 Energies of the parent $H_3C - S - S - (CH_2)_3 - NH_3^+$ cation (open circles), ground Rydberg-attached (open squares), excited Rydberg-attached (filled squares), and S–S σ^*-attached (filed diamonds) states as functions of the S–S bond length. Also shown are the SS σ^* (left), excited-Rydberg (center), and ground-Rydberg (right) orbitals (appears as Figure 4 in ref. 3s).

suggesting that intra-peptide electron transfer can be an important contributor to electrons attaching to and cleaving SS and N–C$_\alpha$ bonds.

In summary, ETD and ECD processes involve two kinds of electron-transfer events. The first occurs in the initial capture of an electron by the positively charged polypeptide. The second involves intra-peptide electron transfer from a Rydberg orbital residing on a positively charged site to an SS or OCN bond site.

2. THE THEORETICAL CHALLENGES AND EXAMPLES OF HOW THE STUDIES ARE PERFORMED

2.1 Theoretical considerations

Before discussing specific examples as a tool for illustrating how one uses theory to carry out such studies, we overview a few components of all theoretical investigations of the electron-transfer events we have studied. Specifically, one must be sure to address all of the following issues:

1. Atomic orbital basis sets containing diffuse functions must be used at least for the atoms onto which the electron will attach. This means the sulfur atoms if one is studying disulfide cleavage and the O, C, and N atoms (at the site of cleavage) if one is studying N–C$_\alpha$ cleavage. It is important to then check to

make sure one obtains a reasonably accurate electron binding energy for the fragment that holds the excess electron upon bond cleavage. For SS bond cleavage, this means verifying that the $^-$S–R anion has an electron binding energy near 1.4 eV. This is important because the relative energies of the bond-attached and Rydberg-attached states determine which Rydberg state is likely to couple to the bond-attached state.

2. The positively charged site to which an electron is to attach must have special basis functions [4–6] attached to it to describe the Rydberg orbitals. This is important because one needs to accurately describe the energies of the Rydberg states in relation to bond-attached states and the Rydberg orbitals' radial extent must be properly represented. To appreciate the sizes of such orbitals, we show in Figure 6 the lowest (labeled 3s, 3p, 3d, 4s, 4p, and 5s because NH_4^+ is isoelectronic with Na^+) Rydberg orbitals of NH_4.

In each orbital, the outer surface in the figure contains only 60% of the electron density (i.e., 40% of the density lies farther from the cation center). Moreover, for each orbital, one can notice the size of the van der Waals surface of the underlying NH_4^+ cation to gain perspective about how large these Rydberg orbitals are. Realizing that the N–H bond length is *ca.* 1 Å, it is easy to appreciate that these Rydberg orbitals span (even at the 60% contour level) 10 Å or more.[2]

3. The theoretical methods used must be capable of describing not only ground but also (several) excited states, including state of the same spatial and spin symmetry. We have found it possible to converge Hartree–Fock self-consistent field (HF-SCF) calculations on excited states by starting the SCF process with a spin-orbital occupancy that describes the desired electronic state. After converging the SCF calculation and checking to make sure it has converged to the correct state, we have employed Møller–Plesset perturbation theory at second order (MP2) to evaluate the energy of each state. A correlated treatment is not so important for the Rydberg-attached states because they

---

[2]Hydrogenic and Rydberg orbitals have "sizes" that can be characterized by their expectation values of r and of r^2:

$$\langle r \rangle_{n,l} = \frac{n^2 a_0}{Z}\left[1.5 - \frac{l(l+1)}{2n^2}\right]; \quad \langle r^2 \rangle_{n,l} = \frac{n^4 a_0^2}{Z^2}\left[2.5 - \frac{3l(l+1)-1}{2n^2}\right]$$

where n and l are the principal and angular momentum quantum numbers of the orbital and a_0 the Bohr unit of length ($a_0 = 0.529$ Å). These expressions can be found, for example, in ref. 7. To conceptualize the magnitude of the overlap (and thus the $H_{1,2}$ coupling strength) of a Rydberg orbital with, for example, a methyl anion lone pair, an SS σ^*, or an amide π^* orbital, think of a Rydberg s-orbital as a spherical shell of radius $\langle r \rangle_{n0} = 1.5n^2 a_0/Z$ having a radial "thickness" δr to its electron distribution characterized by its dispersion in radial distribution $\delta r = [\langle r^2 \rangle_{n,0} - (\langle r \rangle_{n,0})^2]^{1/2} = 0.5n^2 a_0/Z$. This shell of thickness δr thus has a surface area of $4\pi 2.25 n^4 a_0^2/Z^2$ and a volume of $V_n = 4\pi 2.25 \times 0.5n^6 a_0^3/Z^3$. In contrast, a methyl anion lone pair, an SS σ^*, or an amide π^* orbital has a volume of *ca.* $V_{bond} = 4/3\pi(10a_0)^3$. Now, consider one of the latter orbitals penetrating into a Rydberg orbital, and approximate the electron density within each of the two volumes V_n and V_{bond} as uniform. That is, within each volume, the respective wave functions are approximated by $\psi(r) = (1/V)^{1/2}$. The $H_{1,2}$ coupling should then scale with n in the same manner as the overlap integral (S) between the two wave functions $S = \int_{V_{bond}} (1/V_{bond}^{1/2})(1/V_n^{1/2})d^3r = (V_{bond}^{1/2}/V_n^{1/2}) = \sqrt{10^3 Z^3/0.5(3)(2.25)n^6}$ given in terms of the square root of the fraction of the volume of the Rydberg orbital that is shared with the penetrating orbital of volume $(10a_0)^3$. Even for $n = 4$, this overlap is $0.27Z^{2/3}$. For $n = 9$, S is $0.02Z^{3/2}$. This scaling of the overlap between a Rydberg orbital and a valence-sized orbital as n^{-3} suggests that the $H_{1,2}$ couplings will be small except for Rydberg orbitals in the $n = 3$–10 range, not for high-n Rydberg orbitals.

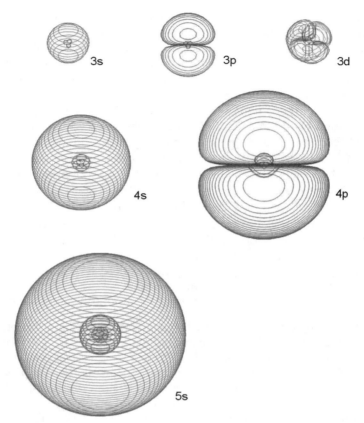

Figure 6 Plots of 3s, 3p, 3d, 4s, and 5s Rydberg orbitals of NH_4 with the outermost contour containing 60% of the electron density of that orbital.

have only one electron in their Rydberg orbital. However, for an anion donor such as H_3C^-, correlation is very important because the extra electron experiences very large correlations with the other methyl lone pair electron.

4. To evaluate the $H_{1,2}$ couplings, one needs to carry out calculations at a very finely spaced grid (often with geometry changes along, for example, the SS bond length, of *ca.* 0.01 Å) in the region of the avoided crossing. After one has determined the smallest energy gap between the two states undergoing the avoided crossing, $H_{1,2}$ is taken an one-half this gap. These same calculations are what one uses to evaluate the slope difference $|\Delta F|$ entering into the LZ surface hopping probability formula.

Finally, it is important to explain the strategy that we have used to construct model compounds on which to carry out *ab initio* calculations from which we can gain insight into the two classes of electron transfer discussed above. For the kind of polypeptides shown in Figures 1 and 4 and for most species used in ETD or ECD experiments, the positively charged sites reside primarily on side chains that possess great motional flexibility. This means that, as the peptide undergoes thermal motion in the gas phase, the distances between the positive sites and any SS or OCN group will fluctuate substantially, as will the distances from one positive site to another. As a result, the Coulomb stabilization energy (Equation (2)) at the SS, OCN, and positive sites will also fluctuate with time. Ideally then, one would like to model the dynamical motions of the polypeptide's side chains and backbone and, at each instant of time, compute the rates for electron transfer from an anion donor to SS, OCN, and Rydberg sites as well as the rates of intra-peptide electron transfer. Such an ideal approach is simply not computationally feasible because of the substantial difficulties involved in each electron transfer rate calculation. Therefore, the approach we have undertaken involves:

a. Using small model compounds containing one disulfide or amide unit to limit the computational cost.
b. Fixing the distances between positive sites and SS or OCN bond sites and between positive sites in each calculation (but varying them from one calculation to another) as a way to gain data representative of that particular set of inter-site distances.

This approach allows us to generate a body of data representative of the range of geometries sampled by a polypeptide undergoing dynamical motions.

2.2 Illustrative examples

With the above advice and strategy in mind, we can now focus on a few illustrative cases involving electron transfer to an SS σ^* orbital that subsequently affects disulfide bond cleavage as a means of further illustrating how these studies proceed and what they have told us. First, let us consider intra-peptide transfer from a Rydberg orbital on a protonated amine site, through intervening aliphatic "spacers" of varying length, to such an SS σ^* orbital.

In Figure 7, we show the SS σ^*, excited-Rydberg, and ground-Rydberg orbitals for three model compounds $^+H_3N-(CH_2)_n-S-S-CH_3$ having $n = 3, 2$, or 1 from left to right.

It is important to recognize that the Rydberg orbitals have significant amplitudes in regions of space where the SS σ^* orbital also does and that the degree of overlap between the Rydberg and SS σ^* orbitals decreases as n increases, as expected.

For $n = 3$, the energy profiles of the parent compound, the species with an electron attached to the ground or excited-Rydberg orbital, and the species with an electron in the SS σ^* orbital as functions of the SS bond length were shown earlier in Figure 3 where we also see the $H_{1,2}$ values associated with the Rydberg SS σ^* avoided crossings. Analogous data was obtained for the $n = 2$ and $n = 3$ cases, and the corresponding $H_{1,2}$ values were obtained. When the ln $H_{1,2}$ values for ground and excited-Rydberg states are plotted for $n = 1, 2$, and 3 are plotted vs. the distance R between the center of the SS bond and the center of charge of the Rydberg orbital, decent linear correlations are obtained as shown in Figure 8.

Such exponential decays of $H_{1,2}$ with distance are characteristic of the electronic coupling strengths in all electron-transfer studies [8–11], not just those

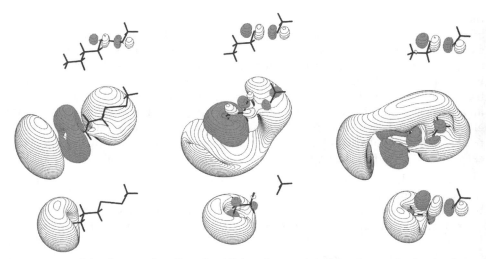

Figure 7 SS σ* (top), excited-Rydberg (middle), and ground-Rydberg (bottom) orbitals of $^+H_3N–(CH_2)_n–S–S–CH_3$ with $n = 3$ (left), 2 (center), and 1 (right) (appears as Figure 5 in ref. 3s).

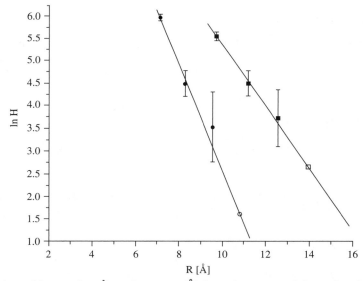

Figure 8 Plots of ln $H_{1,2}$ (cm^{-1}) vs. distance R (Å) from the center of the SS bond and the center of charge of the ground (left line) and excited (right line) Rydberg orbitals for the $^+H_3N–(CH_2)_n–S–S–CH_3$ model compounds having $n = 1$, 2, and 3 (appears as Figure 6 in ref. 3s).

related to intra-peptide or anion-to-peptide electron transfer. The error bars shown in Figure 8 derive from our estimate of how small $H_{1,2}$ can be before we find it too difficult to reliably determine the minimum energy splitting between two surfaces undergoing an avoided crossing.

Although we are not able to directly determine $H_{1,2}$ values as small as $0.3\,\text{cm}^{-1}$ (recall, this is the smallest $H_{1,2}$ that can generate an intra-peptide electron transfer that can compete with relaxations among Rydberg states), we use the near-linear plots of $H_{1,2}$ vs. R to extrapolate to that R-value where $H_{1,2}^{\text{min}} = 0.3\,\text{cm}^{-1}$ should be realized. For example, the data shown in Figure 8 suggest that the excited-Rydberg state can contribute to electron transfer out to $R \approx 18\,\text{Å}$, while the ground-Rydberg state can out to $R \approx 12\,\text{Å}$.

To explore whether the electron-transfer events occur primarily through-space or through-bond, we carried out calculations on model compounds in which the disulfide linkage is separated from the site of the Rydberg orbital(s) by distances similar to those arising in the studies of $^+\text{H}_3\text{N–(CH}_2)_n\text{–S–S–CH}_3$ but with no "spacer" groups between the Rydberg and SS sites. For example, we studied two model systems: $\text{H}_3\text{C–SS–CH}_3$ with an NH_4^+ ion 3–15 Å from the midpoint of the SS bond and $\text{H}_3\text{C–SS–CH}_3$ with an $\text{N(CH}_3)_4^+$ ion 3–15 Å from the midpoint of the SS bond. These two positive sites were chosen to model protonated amine and so-called fixed-charge sites that occur in many polypeptides. The energy profiles of the parent compound and of species with an electron attached to the SS σ^*, ground-, or excited-Rydberg orbitals are shown in Figures 9 and 10.

Also shown in Figures 9 and 10 are the $H_{1,2}$ values (in cm^{-1}) obtained by analyzing the avoided curve crossings. In Figure 11 we show plots of the natural

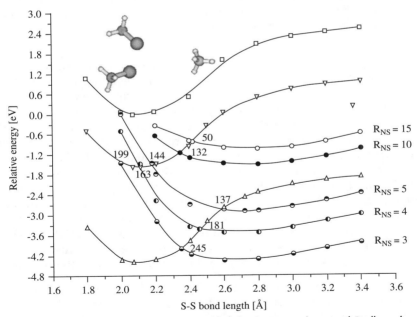

Figure 9 Energies of parent $\text{H}_3\text{C–SS–CH}_3 \ldots \text{NH}_4^+$ (open squares), ground-Rydberg (open triangles), excited-Rydberg (inverted open triangles), and various SS s*-attached (circles) states as functions of the SS bond length, for a range of distances between the nitrogen atom and the midpoint of the SS bond (appears in Figure 8 of ref. 3s).

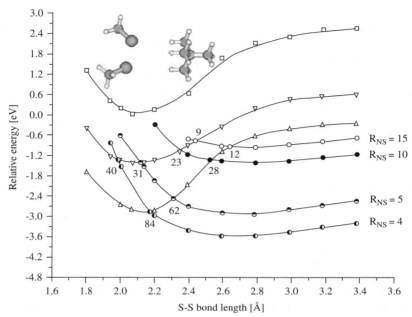

Figure 10 Energies of parent $H_3C-SS-CH_3 \ldots N(CH_3)_4^+$ (open squares), ground-Rydberg (open triangles), excited-Rydberg (inverted open triangles), and various SS s*-attached (circles) states as functions the SS bond length for a range of distances between the nitrogen atom and the midpoint of the SS bond (appears in Figure 8 of ref. 3s).

log of these $H_{1,2}$ values as functions of the distance from the nitrogen atom to the midpoint of the SS bond for the four cases related to Figures 9 and 10.

Again, we see that the Rydberg states' couplings can extend over very large distances. Moreover, it appears (from Figures 8 and 11) that the excited-Rydberg states' coupling strength seems to decay somewhat slower with distance than those of the ground-Rydberg states. Finally, the magnitudes of the $H_{1,2}$ values obtained with $-CH_2-$ spacers present are not qualitatively larger (compare Figures 8 and 11) than those obtained in the through-space study (for a given distance). This suggests that, at least for the systems studied to date, the presence of aliphatic spacers does not qualitatively increase the rates of intra-peptide electron transfer; through-space transfer seems to be dominant.

Although space limitations preclude reviewing all of the results [3h–3u] that have come out of our studies on anion-to-peptide electron transfer and intra-peptide electron transfer, it is worth mentioning here a few of the highlights.

a. In collisions of an anion donor with a positively charged polypeptide, electron transfer to a Rydberg orbital on a positive site is 10–100 times more likely than transfer to an SS σ^* or OCN π^* orbital.

b. Once an electron attaches to a Rydberg orbital (probably an excited orbital), it can relax to lower-energy Rydberg orbitals in *ca.* 1 μs, or it can, in this same timeframe, undergo transfer to any an SS σ^* or OCN π^* orbital that is within

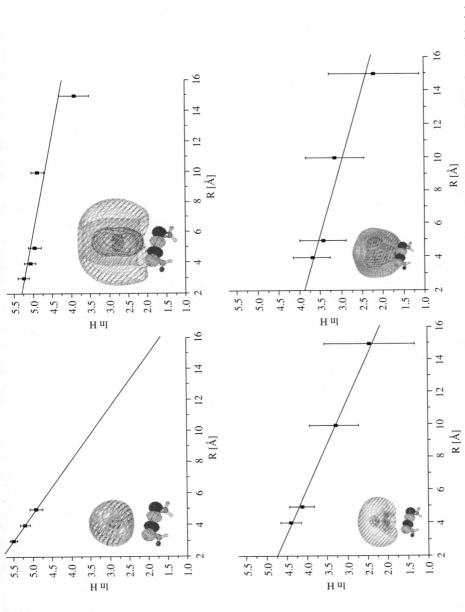

Figure 11 Plots of ln $H_{1,2}$ (cm^{-1}) vs. distance from the nitrogen atom to the midpoint of the SS bond for ground (left) and excited (right) Rydberg states of NH$_4$ (top) and N(CH$_3$)$_4$ (bottom). Also shown are the Rydberg orbitals involved in each case along with the molecular complex's geometry (appears as Figure 7 in ref. 3p).

15–20 Å and that is sufficiently Coulomb stabilized by nearby positive charges to render positive its electron binding energy.

c. Once an electron attaches to a Rydberg orbital, it can transfer to a Rydberg orbital on a different positive site if the two sites come within *ca.* 10 Å of each other.

3. RELATION TO MORE COMMON FORMS OF ELECTRON TRANSFER

Electron-transfer processes play many very important roles in chemistry and biology. Because the present work is focused on electron-transfer events occurring within positively charged gas-phase peptides as they occur in ETD and ECD mass spectrometry experiments, it is not appropriate or feasible to review the myriad of other places electron-transfer reactions occur in chemistry. Chapter 10 of the graduate level textbook by Schatz and Ratner [12] gives a nice introduction to the main kinds of electron-transfer events that chemists usually study as well as to the theoretical underpinnings. They also give, at the end of Chapter 10, several literature references to selected seminal papers on these subjects.

In most other electron-transfer processes, one considers an electron moving from a donor (D) to an acceptor (A) through an intervening molecular structure called a bridge (B). This is much like the Rydberg-bridge-SS system treated earlier in this paper. There are then two diabatic (meaning having a fixed orbital occupancy) electronic states D-B-A and D^+-B-A^- of the donor-bridge-acceptor system between which one views the transfer as taking place. The energy profiles of the reactant (D-B-A) and product (D^+-B-A^-) states as functions of a reaction coordinate X (i.e., the direction along which the two diabatic energy hypersurfaces cross) are, in the most commonly invoked theory, represented as parabolic functions whose minima are shifted in energy by $\varepsilon_2 - \varepsilon_1$ and in length along the reaction coordinate by $X_R - X_L$ as shown in Figure 12.

The two diabatic energy profiles are expressed in terms of harmonic forms having a common force constant as:

$$V_L(X) = \varepsilon_1 + \frac{1}{2}k(X - X_L)^2 \tag{4}$$

$$V_R(X) = \varepsilon_2 + \frac{1}{2}k(X - X_R)^2 \tag{5}$$

The two diabatic surfaces and wave functions are allowed to couple by way of a Hamiltonian matrix element denoted J:

$$J = \langle \psi_L | H | \psi_R \rangle \tag{6}$$

and two adiabatic energy surfaces are generated from the 2×2 Hamiltonian matrix

$$H = \begin{bmatrix} V_L & J \\ J & V_R \end{bmatrix} \tag{7}$$

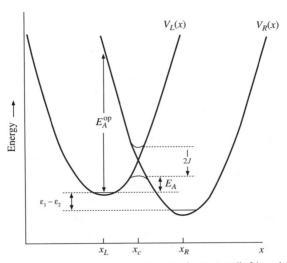

Figure 12 Plots of the energy surfaces appropriate to the D-B-A (left) and D$^+$-B-A$^-$ (right) species as functions of the reaction coordinate along which the diabatic surfaces cross and the adiabatic surfaces undergo an avoided crossing (as shown) (appears as Figure 10.2 in ref. 12).

The two eigenvalues of this matrix

$$E_\pm = \frac{1}{2}[V_L + V_R \pm \sqrt{(V_R - V_L)^2 + 4J^2}]$$

(8)

differ by an amount $2J$ at the point X_C along the reaction coordinate at which the two diabatic curves cross (i.e., $V_L = V_R$ at X_C) as shown in Figure 12. The activation energy E_A (i.e., the energy needed to move from ε_1 to the barrier on the lower adiabatic energy surface (i.e., $E_-(X_C)$)) can be expressed in terms of the so-called reorganization energy Λ and the thermodynamic energy difference $\varepsilon_2 - \varepsilon_1$:

$$E_A = \frac{(\Lambda + \varepsilon_2 - \varepsilon_1)^2}{4\Lambda}$$

(9)

with

$$\Lambda = V_R(X_L) - V_R(X_R)$$

(10)

Λ is called the reorganization energy because (see Figure 12) it is the energy necessary to relax the system when it is in the D$^+$-B-A$^-$ state but at the equilibrium geometry of the D-B-A state (having energy $V_R(X_L)$) to the energy of this D$^+$-B-A$^-$ state at its own equilibrium geometry.

In the cases treated in the present paper, we do not have a reorganization energy because, for example as shown in Figures 5 and 10, the two diabatic states between which electron transfer occurs (e.g., the SS σ^* and excited-Rydberg states) cross so close (i.e., within the zero-point vibrational motion of the SS bond) to the minimum on the Rydberg-state surface as to render Λ essentially zero. In more traditional electron-transfer events, Λ contains contributions from the energy needed to rearrange the geometry of the D-B-A molecule itself as well as the energy needed to relax the surrounding solvent environment to the change from D-B-A to D$^+$-B-A$^-$. That is, in D-B-A the surrounding solvent experiences a very different electrostatic potential than in D$^+$-B-A$^-$, so the solvent molecules must reorient (and polarize) to adjust to the change in this potential. However, as noted above, in our case, there is no intramolecular reorganization energy and no solvent contribution because the mass spectroscopy experiments are carried out in the gas phase.

Returning to the more common electron-transfer cases, as shown in ref. 12, the electron-transfer rate is eventually expressed as a product of two terms. One term, which depends on the activation energy E_A in the usual $\exp(-E_A/RT)$ manner contains the reorganization energy. The other term is proportional to J^2 and reflects the intrinsic electron-transfer rate once the system reaches the activation barrier. The scaling with J^2 arises when the couplings between the two diabatic states are treated perturbatively in this so-called nonadiabatic limit. In the cases treated in this paper, the electron-transfer rates depend on $H_{1,2}^2$ ($H_{1,2}$ is the same as J) through the LZ expression, but we have no $\exp(-E_A/RT)$ factor because, as already explained, our reorganization energies are essentially zero. They scale as $H_{1,2}^2$ because, in the LZ estimate of the surface hopping probability, the two diabatic states that cross are assumed to undergo a weakly avoided crossing; that is, the LZ estimate is in line with the nonadiabatic limit discussed in conventional electron-transfer theory.

Finally, it may be useful to note that the Fermi golden rule and time correlation function expressions often used (see ref. 12, for example) to express the rates of electron transfer have been shown [13], for other classes of dynamical processes, to be equivalent to LZ estimates of these same rates. So, it should not be surprising that our approach, in which we focus on events with no reorganization energy requirement and we use LZ theory to evaluate the intrinsic rates, is closely related to the more common approach used to treat electron transfer in condensed media where the reorganization energy plays a central role in determining the rates but the J^2 factor plays a second central role.

In closing, it may be instructive to contrast the electron-transfer events taking place in polypeptides with those we have been studying relating to electrons in DNA [14]. In these studies, we simulate processes in which

a. an electron attaches to a π^* orbital on one of DNA's bases, after which
b. the electron can autodetach, or
c. it can undergo a transfer through the sugar unit attached to the base and into the sugar-phosphate C–O σ bond's antibonding orbital, thus leading to C–O bond cleavage and a so-called single strand break.

The branching ratio between autodetachment and electron transfer governs the yield of strand breaks. In Figure 13, we show a qualitative depiction of the energy surfaces involved in this class of electron-transfer processes.

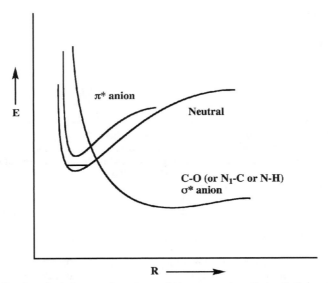

Figure 13   Qualitative depiction, as functions of the sugar-phosphate C–O bond length, of the energy of a base-sugar-phosphate nucleotide with no electron attached (labeled neutral), with an electron attached to its base π^* orbital (labeled π^* anion), and with the electron residing in the sugar-phosphate C–O σ^* orbital (lower curve) (appears as Figure 7 in ref. 14).

There are two primary differences in this DNA case when compared to the polypeptide systems discussed earlier:

1. Because the repulsive C–O σ^*-attached state crosses the base π^*-attached state at an energy significantly above the minimum on the π^*-attached state's surface (see Figure 13), the C–O bond must undergo substantial elongation to access this crossing point. This elongation is thought to occur by thermal excitation of the C–O stretching motion. The energy ΔE required to reach this crossing is analogous to the reorganization energy discussed earlier. This requirement gives rise to a Boltzmann $\exp(-\Delta E/RT)$ dependence in the electron-transfer rate for this DNA case, much like the reorganization energy does in the conventional electron-transfer theory discussed earlier.

2. The $H_{1,2}$ matrix elements connecting the C–O σ^*-attached and the base π^*-attached states were found [14] to be much larger (e.g., $>1{,}000\,\mathrm{cm}^{-1}$) than in the polypeptide case (where they were usually $<300\,\mathrm{cm}^{-1}$). As a result, the DNA electron transfer does not occur in the nonadiabatic limit discussed earlier as it does in the polypeptides. In the DNA case, the couplings are large enough that the system evolves adiabatically (i.e., once the barrier at the crossing of the C–O σ^*-attached and the base π^*-attached states is reached, electron transfer is prompt) from the base to the sugar-phosphate C–O bond that is then cleaved.

ACKNOWLEDGMENT

This work has been supported by NSF Grant No. 0806160.

REFERENCES

1. [a] Zubarev, R.A., Kelleher, N.L., McLafferty, F.W. Electron capture dissociation of multiply charged protein cations. A nonergodic process. J. Am. Chem. Soc. 1998, 120, 3265–6. [b] Zubarev, R.A., Kruger, N.A., Fridriksson, E.K., Lewis, M.A., Horn, D.M., Carpenter, B.K., McLafferty, F.W. Electron capture dissociation of gaseous multiply-charged proteins is favored at disulfide bonds and other sites of high hydrogen atom affinity. J. Am. Chem. Soc. 1999, 121, 2857–62. [c] Zubarev, R.A., Horn, D.M., Fridriksson, E.K., Kelleher, N.L., Kruger, N.A., Lewis, M.A., Carpenter, B.K., McLafferty, F.W. Electron capture dissociation for structural characterization of multiply charged protein cations. Anal. Chem. 2000, 72, 563–73. [d] Zubarev, R.A., Haselmann, K.F., Budnik, B., Kjeldsen, F., Jensen, F. Account: towards an understanding of the mechanism of electron-capture dissociation: a historical perspective and modern ideas. Eur. J. Mass Spectrom. 2002, 8, 337–49.
2. [a] Syka, J.E.P., Coon, J.J., Schroeder, M.J., Shabanowitz, J., Hunt, D.F. A bubble-driven microfluidic transport element for bioengineering. Proc. Natl. Acad. Sci. 2004, 101, 9523–8. [b] Coon, J.J., Syka, J.E.P., Schwartz, J.C., Shabanowitz, J., Hunt, D.F. Anion dependence in the partitioning between proton and electron transfer in ion/ion reactions. Int. J. Mass Spectrom. 2004, 236, 33–42. [c] Pitteri, S.J., Chrisman, P.A., McLuckey, S.A. Electron-transfer ion/ion reactions of doubly protonated peptides: effect of elevated bath gas temperature. Anal. Chem. 2005, 77, 5662–9. [d] Gunawardena, H.P., He, M., Chrisman, P.A., Pitteri, S.J., Hogan, J.M., Hodges, B.D.M., McLuckey, S.A. Electron transfer versus proton transfer in gas-phase ion/ion reactions of polyprotonated peptides. J. Am. Chem. Soc. 2005, 127, 12627–39. [e] Gunawardena, H.P., Gorenstein, L., Erickson, D.E., Xia, Y., McLuckey, S.A. Electron transfer dissociation of multiply protonated and fixed charge disulfide linked polypeptides. Int. J. Mass Spectrom. 2007, 265, 130–8.
3. [a] Syrstad, E.A., Turecek, F. Hydrogen atom adducts to the amide bond. Generation and energetics of the amino(hydroxy)methyl radical in the gas phase. J. Phys. Chem. A 2001, 105, 11144–55. [b] Turecek, F., Syrstad, E.A. Mechanism and energetics of intramolecular hydrogen transfer in amide and peptide radicals and cation-radicals. J. Am. Chem. Soc. 2003, 125, 3353–69. [c] Turecek, F., Polasek, M., Frank, A., Sadilek, M. Transient hydrogen atom adducts to disulfides. Formation and energetics. J. Am. Chem. Soc. 2000, 122, 2361–70. [d] Syrstad, E.A., Stephens, D.D., Turecek, F. Hydrogen atom adducts to the amide bond. Generation and energetics of amide radicals in the gas phase. J. Phys. Chem. A 2003, 107, 115–26. [e] Turecek, F. NCα bond dissociation energies and kinetics in amide and peptide radicals. Is the dissociation a non-ergodic process? J. Am. Chem. Soc. 2003, 125, 5954–63. [f] Syrstad, E.A., Turecek, F. Toward a general mechanism of electron capture dissociation. J. Am. Soc. Mass Spectrom. 2005, 16, 208–24. [g] Uggerud, E. Electron capture dissociation of the disulfide bond—a quantum chemical model study. Int. J. Mass Spectrom. 2004, 234, 45–50. [h] Anusiewicz, I., Berdys-Kochanska, J., Simons, J. Electron attachment step in electron capture dissociation (ECD) and electron transfer dissociation (ETD). J. Phys. Chem. A 2005, 109, 5801–13. [i] Anusiewicz, I., Berdys-Kochanska, J., Skurski, P., Simons, J. Simulating electron transfer attachment to a positively charged model peptide. J. Phys. Chem. A 2006, 110, 1261–6. [j] Sawicka, A., Skurski, P., Hudgins, R.R., Simons, J. Model calculations relevant to disulfide bond cleavage via electron capture influenced by positively charged groups. J. Phys. Chem. B 2003, 107, 13505–11. [k] Sobczyk, M., Skurski, P., Simons, J. Dissociative low-energy electron attachment to the C–S bond of H3C-SCH3 influenced by Coulomb stabilization. Adv. Quantum Chem. 2005, 48, 239–51. [l] Sawicka, A., Berdys-Kochaska, J., Skurski, P., Simons, J. Low-energy (0.1 eV) electron attachment S–S bond cleavage assisted by Coulomb stabilization. Int. J. Quantum Chem. 2005, 102, 838–46. [m] Anusiewicz, I., Berdys, J., Sobczyk, M., Sawicka, A., Skurski, P., Simons, J. Coulomb-assisted dissociative electron attachment: application to a model peptide. J. Phys. Chem. A 2005, 109, 250–8. [n] Bakken, V., Helgaker, T., Uggerud, E. Models of fragmentations induced by electron attachment to protonated peptides. Eur. J. Mass Spectrom.

2004, 10, 625–38. [o] Skurski, P., Sobczyk, M., Jakowski, J., Simons, J. Possible mechanisms for protecting N–Ca bonds in helical peptides from electron-captue (or transfer) dissociation. Int. J. Mass Spectrom. 2007, 265, 197–212. [p] Sobczyk, M., Neff, D., Simons, J. Theoretical study of through-space and through-bond electron transfer within positively charged peptides in the gas phase. Int. J. Mass Spectrom. 2008, 269, 149–64. [q] Sobczyk, M., Simons, J. Distance dependence of through-bond electron transfer rates in electron-capture and electron-transfer dissociation. Int. J. Mass Spectrom. 2006, 253, 274–80. [r] Sobczyk, M., Simons, J. The role of excited Rydberg states in electron transfer dissociation. J. Phys. Chem. B 2006, 110, 7519–27. [s] Neff, D., Sobczyk, M., Simons, J. Through-space and through-bond electron transfer within positively charged peptides in the gas phase. Int. J. Mass Spectrom. 2008, 276, 91–101. [t] Neff, D., Simons, J. Theoretical study of electron capture dissociation of $[Mg(H_2O)_n]^{2+}$ clusters. Int. J. Mass Spectrom. 2008, 277, 166–74. [u] Simons, J. Molecular anions. J. Phys. Chem. A 2008, 112, 6401–511. [v] Neff, D., Smuczynska, S., Simons, J. Electron shuttling in electron transfer dissociation. Inter. J. Mass Spec. 2009, 283, 122–34. [w] Neff, D., Simons, J. Analytical and computational studies of intra-molecular electron transfer pertinent to electron transfer and electron capture dissociation mass spectrometry. J. Phys. Chem. A (submitted, 2009).

4. Gutowski, M., Simons, J. Lifetimes of electronically metastable double-Rydberg anions: FH_2^-. J. Chem. Phys. 1990, 93, 3874–80.

5. Simons, J., Gutowski, M. Double-Rydberg molecular anions. Chem. Rev. 1991, 91, 669–77.

6. Skurski, P., Gutowski, M., Simons, J. How to choose a one-electron basis set to reliably describe a dipole-bound anion. Int. J. Quantum Chem. 2000, 80, 1024–38.

7. Pauling, L., Wilson, E.B. Jr. Introduction to Quantum Mechanics with Applications to Chemistry, Dover Publications, New York, 1985.

8. McConnell, H.M. Intramolecular charge transfer in aromatic free radicals. J. Chem. Phys. 1961, 35, 508.

9. Mujica, V., Kemp, M., Ratner, M.A. Electron conduction in molecular wires. II. Application to scanning tunneling microscopy. J. Chem. Phys. 1994, 101, 6856.

10. Jordan, K.D., Paddon-Row, M.N. Long-range interactions in a series of rigid nonconjugated dienes. 1. Distance dependence of the π_+, π_- and π_+^*, π_-^* splittings determined by ab initio calculations. J. Phys. Chem. 1992, 96, 1188.

11. Curtiss, L.A., Naleway, C.A., Miller, J.R. Superexchange pathway calculation of long-distance electronic coupling in $H_2C(CH_2)_{m-2}CH_2$ chains. Chem. Phys. 1993, 176, 387.

12. Schatz, G.C., Ratner, M.A. Quantum Mechanics in Chemistry, Prentice Hall, Englewood Cliffs, NJ, 1993.

13. Taylor, H., Simons, J. A different view of molecular electronic transitions. J. Phys. Chem. 1986, 90, 580–3.

14. Simons, J. How do low-energy (0.1–2 eV) electrons cause DNA strand breaks? Acc. Chem. Res. 2006, 39, 772–9.